TEACHING ADULTS

A Math Resource Book

Libby Serkies, M.Ed.
Firelight Education

Reviewer:
Kathryn B. Thompson
Executive Director
Tarrant Literacy Coalition
Fort Worth, Texas

New Readers Press
ProLiteracy's publishing division

Acknowledgments

To Derek, Tristan, Jax, and Connor: my hearts! I love you all so much! To Mom who is my number one fan, and to Dad watching in Heaven.

To my Adult Numeracy Network colleagues: Cynthia Bell, Donna Curry, Lynda Ginsburg, Margaret Giordano, Brooke Istas, Pam Meader, Connie Rivera, and Sally Waldron—what an incredible group of passionate and inspiring math educators you are! I have learned so much from all of you!

To my Minnesota friends and colleagues: Amber Delliger, Brad Hasskamp, Astrid Liden, Abby Roza, Rebecca Strom, and Amy Vickers (even though you live in Wisconsin now)—I have the most fun learning and planning with you. You are all rock stars!

To Cindy Tesdall—supervising teacher extraordinaire and where it all started!

To Dannie Francis—the best training partner I have ever had!

And finally, to Crabby: You're my favorite crustacean in the whole world!

Any inaccuracies contained within this work are wholly mine. –Libby

Teaching Adults: A Math Resource Book
ISBN 978-1-56420-878-1

Copyright © 2016 New Readers Press
New Readers Press
ProLiteracy's Publishing Division
104 Marcellus Street, Syracuse, New York 13204
www.newreaderspress.com

Printed in the United States of America
9 8 7 6 5 4 3 2 1

Proceeds from the sale of New Readers Press materials support professional development, training, and technical assistance programs of ProLiteracy that benefit local literacy programs in the U.S. and around the globe.

Writer: Libby Serkies
Editorial Director: Terrie Lipke
Assistant Production Editor: Laura McLoughlin
Illustrations: Jamie Wallace
Technology Specialist: Maryellen Casey
Cover Designer: Carolyn Wallace

Contents

Video Activities

Transitioning from Concrete to Representational to Abstract (p. 21)

Making and Using Fraction Packets (p. 28)

Solving Relevant Word Problems (p. 29)

Operations with the Place Value Mat and Chips (p. 88)

Operations on an Open Number Line (p. 93)

Using Ten Frames (p. 93)

Using Base Ten Blocks (p. 93)

Multiplication Strategies (p. 100)

Division Strategies (p. 101)

Fraction Operations (p. 123)

Using Algebra Tiles (p. 143)

Geometry with Pattern Blocks (p. 166)

Introduction

This Book

Teaching Adults: A Math Resource Book is designed to provide tutors and instructors with tools to help them improve their students' math skills. It incorporates some of the most effective instructional strategies that have proven successful with adults. It also includes some background information on how the field of mathematics has evolved and where it is today.

Do you or your students suffer from math anxiety? Are you a volunteer tutor? Are you an instructor with little or no experience teaching math? Don't let your anxiety keep you from helping your students, and don't let your students' anxiety keep them from learning. This book will give you strategies to successfully navigate math instruction without anxiety—yours or your students'.

This resource book will help you to

- Understand the shift to more rigorous math standards and expectations
- Discover the *why before how* principle—and why it helps adult learners
- Locate and fill the gaps in students' mathematical competencies
- Provide opportunities for students to succeed with math
- Present math concepts in a variety of ways
- Learn to use manipulatives to present complex concepts in a concrete way

In addition to the activities printed in this book, you will see references to video activities. To view the video activities, go to http://www.newreaderspress.com/teaching-adults-math-resource-book. Click on Free Resources to access the video link.

Math Instruction

Why Math Instruction Is Different Now

Worldwide, education is constantly changing. And you may have noticed that mathematics instruction has been undergoing a shift in recent years. In the United States, the shift has had a profound impact.

Students in the K-12 school system probably don't notice the changes. But some instructors who were trained decades ago may struggle with the way skills and concepts are presented today. Students are learning new concepts or learning them at an earlier age. Parents are having trouble helping their kids with math homework because operations are taught differently than they were 10, 20, or 30 years ago. There is less memorization of facts and more emphasis on problem-solving strategies.

These changes began to affect the adult education field at about the same time as K-12 schools—when the Common Core State Standards (CCSS) were released in 2010 as the benchmark for academic achievement in many states. In states where adult education programs are offered through local school districts, you may have already been planning instruction around these standards for years.

The College and Career Readiness Standards (CCRS) for adult education were actually developed first, but they were not introduced until after the K-12 standards were in use. With the long delay in the reauthorization of the Workforce Investment Act (WIA)—now the Workforce Innovation and Opportunity Act (WIOA)—there had been no real impetus to change adult education until the GED® test was revised in January 2014.

It is okay to feel intimidated by what seems like a dramatic shift in math education. That shift is real. There are a number of factors that have contributed to the shift in math instruction, and many will be discussed in this book.

Why Math Had to Change

Mathematical standards-based educational reform had to happen. For decades passing the GED® test was considered not only as an equivalent to achieving a high school diploma, but also as signaling that recipients were ready to enter college or the workplace. As time went on, this became further and further from reality. Preparing adult learners to be successful when pursuing a professional certification, a two- or four-year degree, a promotion, or a family-sustaining wage has always been our goal. However, the 2002 GED® credential was no longer enough.

The new standards and the new high school equivalency tests that followed in 2014—GED®, TASC, and HiSET®—were developed to raise the bar and ensure students really are prepared. New math questions reflect increased rigor, higher standards, and even include technology on the computer-based tests. Before the 2014 GED® test was even released, people were talking about how much harder it was going to be than the 2002 series. These were some of the new terms we heard:

- Rigor
- Complexity
- Webb's Depth of Knowledge
- Algebraic Reasoning
- Quantitative Reasoning

Clearly the test developers were digging deep into the new standards and looking at math in a new way.

How the Changes Affect You and Your Students

Now you are charged with preparing students to step up and rise to these higher standards. To do that, you may want to look back and examine what has historically been done in terms of math instruction. Then you can reflect on what worked, what did not work, why things failed, and most importantly, learn the unknowns about teaching adults math. Those unknowns include the previous experiences students have had with math, the gaps in students' conceptual understanding, and the tools and strategies it will take to fill in those gaps.

Occasionally students have gaps in their mathematical competencies, and it's your job to help fill those gaps. In the book *Why Before How* (2011), Jana Hazekamp emphasizes that, as an instructor, you need to present students with opportunities to experience math, instead of just showing them an algorithm to memorize and then giving them practice problems to complete. Consider the goal of helping students to achieve procedural fluency:

> Procedural fluency is the ability to apply procedures accurately, efficiently, and flexibly; to transfer procedures to different problems and contexts; to build or modify procedures from other procedures; and to recognize when one strategy or procedure is more appropriate to apply than another. (National Council of Teachers of Mathematics (NCTM))

Procedural fluency is a worthy goal, but building a solid foundation on conceptual understanding is crucial to future success in mathematics. Why can't you just tell students $9 + 7$ is 16? Some students get it; they understand with no difficulty that $9 + 7 = 16$. But for those who do not get it, and who need things broken down completely, you need to be able to do that for them.

Finding and filling in the gaps in students' conceptual understanding of math is *crucial* to their future success. For instance, consider the following misconceptions and how they will hamper students' progress in math:

- Quantities: 0.5 is less than 0.25.
  (Incorrect reasoning: 5 is less than 25.)

- Quantities: A three-digit number is always smaller than a four-digit number.
  (Incorrect reasoning: 2.65 is smaller than 1.096.)

- Fractions/Quantities: The fraction ⅓ is less than ⅕.
  (Incorrect reasoning: 3 is less than 5.)

- Base ten: To multiply by 10, just add a 0.
  (Incorrect reasoning: what if you are multiplying a decimal by 10?)

- Proportion: If 4 out of a dozen donuts are raspberry filled, and 3 are strawberry filled, what proportion of the donuts are fruit filled?
  (Incorrect reasoning: many students will say ¾, comparing strawberry to raspberry, instead of comparing the total of fruit filled donuts to the total number of donuts.)

- Operations: A negative *and* a negative is always a positive.
  (Incorrect reasoning: this is true in multiplication, but students like to apply this concept to lots of other situations where it may not be true.)

If students have holes like these in their mathematical foundations, nothing built on them will be stable. This also ties into procedural fluency, another key component of mathematical success. If students are able to apply known patterns or concepts to a new problem, select the best problem solving strategy, and solve the problem accurately and efficiently, their procedural fluencies will improve dramatically.

Negative Mindsets

As adult learners gain confidence in their abilities to understand and apply their skills to new math, they become more likely to take risks and attempt solutions instead of engaging in defeating self-talk. Most of your students have probably, at one time or another, said or thought something like one of these:

- I've never been good at math.

- I can't do math.

- No one in my family is good at math.

- I failed this before; I will fail it again.

- I am just stupid where math is concerned.

- I can't.

- I won't.

- I'm a failure.

It takes more than a willing spirit for students who have such negative thoughts to begin believing they can be successful in a math classroom or on a math test. And with all this echoing in their minds, how could we ever expect them to be willing? It's not just about filling in the gaps in their understanding, it's about creating new experiences for them. A professor of mathematics education at Stanford University, Dr. Jo Boaler, published a new book called *Mathematical Mindsets*. In it, Dr. Boaler describes the differences between fixed and growth mindsets, and how those differences can profoundly impact the success a student has in class. A very rudimentary explanation of the mindsets goes like this: In fixed mindsets, people believe that they learn because they are smart, or they have the right genetic makeup that makes learning easy. In growth mindsets, people believe that they learn with hard work and persistence.

Dr. Boaler believes "there is no such thing as a 'math brain' or a 'math gift,'" (Boaler, 2016). The interesting thing is many math professors believe there is such a thing! When I was in my undergrad studies, although most math topics came relatively easy to me, I struggled mightily with calculus. As a matter of fact, my calculus professor told me after I bombed my first exam that I couldn't "differentiate my way out of a paper bag." If I had had a fixed mindset, if I thought I was gifted with the math gene, I might have stopped there. I might have thought I had just been lucky with the other math topics, and now that it was getting harder, I had reached my limit. But I was lucky in a different way: I believed that hard work would pay off.

Dr. Boaler writes, "It is imperative for our society that we move to a more equitable and informed view of mathematics learning in our conversations and work with students" (Boaler, 2016). Dr. Boaler and a colleague analyzed the test scores of students from around the world who took the Programme for International Student Assessment (PISA) exams. They also looked at the survey results for the students' ideas and beliefs about math and their mindsets. The results showed that "the highest achieving students in the world are those with a growth mindset, and they outrank the other students by the equivalent of more than a year of mathematics" (Boaler, 2016). I encourage you to read more about growth mindsets. For me, as a teacher, the implications are many. They include adding the following ideas to my math classes:

- Ask students what they believe about math success and failure; engage the class in a discussion

- Normalize fears (no one wants to fail in front of an audience)

- Create a safe space for your students in which to learn and support each other

- Refrain from using labels (i.e., smart, talented, gifted, etc.)

- Provide opportunities for students to begin having success with math

Use this information to be conscious not only of what and how you teach, but the words you use, and the messages you send as you interact with your students.

3 Instructional Models

Along with gaps in their mathematical competencies, students may not have enough tools in their problem-solving toolbox. One of the best things about teaching math is that you can often use more than one way to solve a problem. This gives you freedom and creativity to introduce processes that students may not have seen before. Providing students with more tools to work with means that they will be better equipped to persist in solving problems.

Think outside the box when you try to find ways to help your students understand math. For example, do you have a student who loves music but just cannot understand how to work with fractions? Why not use musical notes and staff paper? Show him how fractions represent whole notes, half notes, and eighth notes. Have him practice writing the fractions, adding and subtracting the fractions, and comparing the fractions using their musical note equivalents.

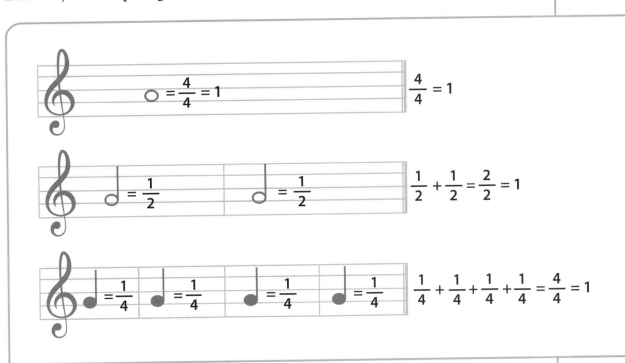

Many math classes beyond the primary level follow the same format: lecture on a concept, show some examples, assign some practice problems as homework, and repeat daily. I have always been a very visual learner, and I often had to find my own way to understand the instruction by creating and manipulating drawings. A lot of students have not tried that or might not realize it's even an option. Now that these students are adults, it's up to us to present math instruction in whatever way will help them to make sense of it. That also means that when we ask them to demonstrate their understanding of a concept that we are open to letting them show or describe it in whatever way works for them.

As a visual learner, I really like using pictures and drawings as instructional tools. Let's face it, we teach the way we like to learn. Even if you don't prefer visual representations, you may have some students who do. You may also have students who are kinesthetic learners and learn best when they can get their hands on manipulative tools. The activities in this book will model several teaching and learning styles. If you can get used to presenting math concepts in a variety of ways, you increase the chances of reaching your students.

Math instruction is compatible with and adaptable to a variety of learning styles, such as auditory, visual, kinesthetic, or even a combination of those learning styles. You may already be familiar with preferred learning styles, but if you would like to read more about this topic, refer to Appendix G.

Start with *Why* Before *How*

One instructional model to consider is the Singapore Math® model. Singapore Math (SM) focuses on building fundamental math skills to give students a solid foundation for math learning. It has three main tenets:

1. Developing number sense

2. Applying model drawing

3. Practicing mental math

The SM model encourages you to move "away from an emphasis on teaching computations as series of rote rules" (Hazekamp, 2011). Instead of demonstrating a formula and then handing out worksheets for practice, with SM you might begin class by using blocks or drawings to make sure students can demonstrate concrete understanding of a math concept before you move to more abstract practice problems.

"SM focuses on relational understanding of mathematics versus instrumental understanding" (Hazekamp, 2011). Instrumental understanding is what Hazekamp refers to as "the ability to follow rote rules." It's the old-fashioned "drill and kill" method of making students mindlessly practice the same skill over and over again. But with SM instruction, the focus is on relational understanding. Students do not practice procedural fluency until they truly understand the mathematical concepts and reasoning underlying the rules.

You may have no problem explaining the *how* in math class, but struggle to clearly explain the *why*, and that can make it difficult for students to integrate the instruction. For example, to explain how to subtract 9 from 12, your words need to accurately describe what is happening mathematically. There are students who might have difficulty conceptualizing the process when you use vague terms such as *borrowing* and *carrying*. So you might need to find a different, more accurate framework altogether.

Conceptual understanding is integral to the CCRS, and to the high school equivalency tests. It's one of the strands of mathematical proficiency specified in the National Research Council's report *Adding It Up: Helping Children Learn Mathematics* (2001).

Strands of Mathematical Proficiency

- adaptive reasoning

- strategic competence

- conceptual understanding (comprehension of mathematical concepts, operations, and relations)

- procedural fluency (skill in carrying out procedures flexibly, accurately, efficiently, and appropriately)

- productive disposition (habitual inclination to see mathematics as sensible, useful, and worthwhile, coupled with a belief in diligence and one's own efficacy)

Often, it is those students who struggle with the concepts that force us to look for new ways to explain the *why*, while at the same time re-evaluating methods to teach the *how*.

If a student struggles with the concepts of *borrowing* and *carrying* as computational strategies, try using other terms like *breaking apart* and *putting together* to describe what is happening. These terms more accurately describe the process. Sometimes you get so used to using certain terms that you don't realize students may not understand what they mean.

The traditional model or algorithm for subtraction looks something like this:

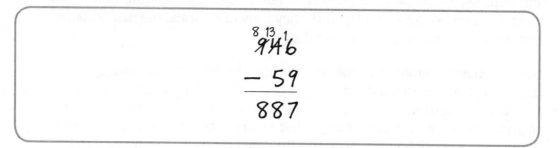

In this example, notice the "borrow and carry" method. There aren't enough ones (6) from which to subtract 9. It also lacks a sufficient number of tens (40) from which to subtract 50. In the first step cross out the 4, write a 3 above it, and put a 1 near the 6 to indicate it is now 16. Perform the calculation and write a 7 below the total line. Repeat the process to make enough tens from which to subtract 5. The concept behind the process is correct, however the method often leaves students confused. When you borrow and carry, you are really regrouping the numbers to make the computation easier. Here is a way to demonstrate regrouping using a place value mat and chips:

First, represent the minuend (the first number) of 946 using 9 hundreds, 4 tens, and 6 ones chips.

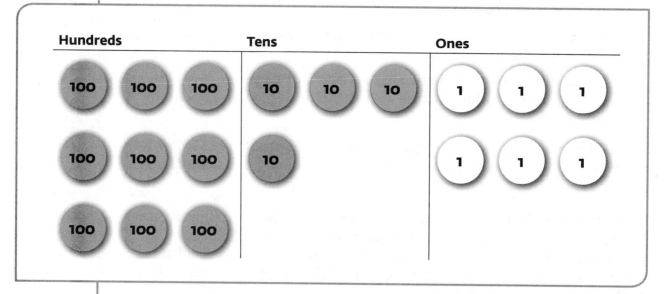

Because there are not enough ones or tens in the chart to perform the operation, you need to decompose the tens and hundreds and regroup them in the appropriate columns. Decompose 1 ten into 10 ones. Likewise, 1 hundred is decomposed into 10 tens. The chips are then regrouped into the ones and tens columns, respectively. Now there are enough chips in each column to perform the calculation.

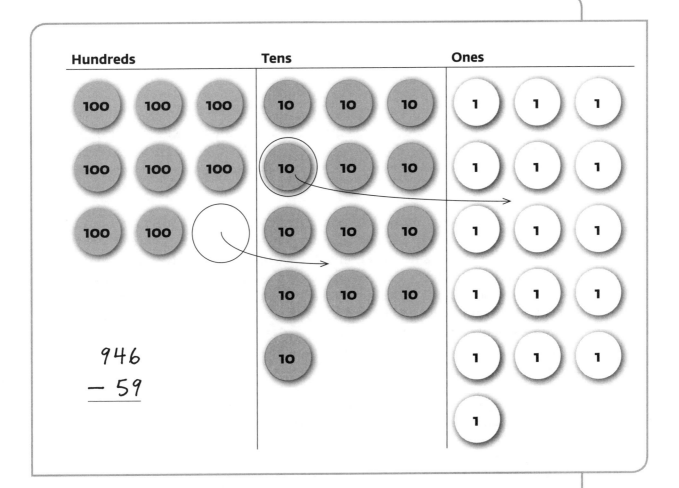

$$946$$
$$-\ 59$$

Next, perform the calculation: Subtract 9 ones and 5 tens from the minuend.

To finish the problem, regroup the digits.

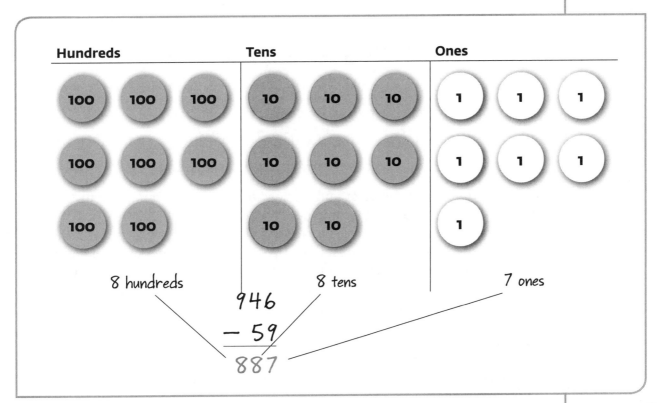

8 hundreds   8 tens   7 ones

$$946$$
$$-\ 59$$
$$\overline{887}$$

It's important that students thoroughly understand the foundational concepts of math, such as how to perform computations, before introducing shortcuts or new content. Since math concepts build on each other, if you move on too quickly, students may have trouble learning new concepts or skills.

Do you ever feel like you have too little time with your students and too much to do to get them ready for college or work? You spend so much class time trying to figure out what they don't know, that students get frustrated and leave. Can you relate to that?

Sometimes students are anxious to move on. They are often working toward an important goal, such as high school equivalency, and they may worry that they are not progressing quickly enough. It's up to you to guide and encourage students to move at the proper pace. Using the place value mat and chips may seem cumbersome to you or it may seem like too many steps. But it's important to take the time to get the foundational skills right. If your students have gaps in this particular area, using this instructional strategy will fill them. After using the place value mat, students may understand how the shortcut steps in the borrow-and-carry or regrouping method work.

Try to keep your students motivated by first acknowledging what they have done. They took the brave step of going back to school, so find out what brought them back. Find out what motivates them, and use that to keep them going. Also—this is important—be willing to step outside your comfort zone and explore new ways of delivering math instruction. This flexibility in presenting lessons is crucial to the shift in standards-based math instruction. To that end, let's look at the topic of Universal Design.

Universal Design for Learning

There are three primary principles of Universal Design for Learning (UDL):

- There are multiple means of representing content
- There are multiple means of action and representation for students
- There are multiple means of engagement in learning

All adult students are not alike, and we need to be able to differentiate instruction for students with different learning needs. However, sometimes we get complacent, stuck in a rut, or tied to a specific method of assessment, whether it be worksheets, quizzes, posttests, etc. UDL stresses the importance of having several methods of instruction and several strategies to elicit engagement. Every student is unique, and we need to have a variety of instruction, assessment, and engagement tools in order to reach each one.

You will read more about various ways to present instruction later in this chapter. The video activities included with this book will demonstrate alternative teaching methods. For more information on UDL, see the resource section in Appendix G.

Mathematical Reasoning

When the 2014 GED® test was being prepared, we heard that the math test would be computerized and more rigorous. We also learned that the math portion of the 2014 test would be called "Mathematical Reasoning." Obviously this would not be a test asking students to apply an algorithm to 30 multiple-choice problems.

How often do students say they just want to know the formula to solve the problem? In his 2010 TED Talk "Math Needs a Makeover," Dan Meyer said "there are five symptoms you're doing math reasoning wrong:"

1. Lack of initiative

2. Lack of perseverance

3. Lack of retention

4. Aversion to word problems

5. Eagerness for a formula

Do any of these sound familiar? They all describe how students might feel when finishing a unit or lesson. As Meyer says, "you finish your lesson, and immediately, you have a dozen hands go up, asking you to start all over with your instruction again." You can avoid this outcome by making some subtle but profound changes in your instruction.

In his TED Talk, Meyer goes on to say that when looking at most textbooks, "what we are doing is taking a compelling question, a compelling answer, but we're paving a smooth, straight path from one to the other, and congratulating our students for how well they can step over the small cracks in the way."

Think about your lessons and begin framing them by asking: *How* will the problems and activities I use in my instruction:

1. Stretch my students' abilities by asking them to apply previous content to a new problem?

2. Encourage the development of reasoning and problem solving skills while building conceptual mastery?

Time spent being purposeful about the tasks and activities you ask students to do should yield results in the form of building their competencies in mathematics. Doing this will help the landscape of math instruction—and hopefully math instructor training—to shift to a focus on helping students to not just "do math," but to learn, apply, and understand math.

See Appendix G for a link to Meyer's TED Talk.

The CRA Model: Concrete, Representational, and Abstract

An effective instructional model for math comprehension is the CRA model—Concrete, Representational, and Abstract. This model helps create real connections to mathematical concepts and incorporates the three basic learning preferences: auditory, visual, and kinesthetic. This is an effective model whether beginning instruction in a new concept, or trying to address the gaps in a student's math education.

The CRA model is a three-step instructional approach:

1. Concrete: teach the math concept using manipulatives

2. Representational: introduce images to represent the concept

3. Abstract: show the concept using numbers and symbols

However, according to Ma (1999), not all manipulative lessons are the same, and can be harmful to early math learning if applied inappropriately:

> Most of the U.S. teachers said they would use manipulatives to help students understand the fact that 1 ten equals 10 ones. In their view, of the two key steps of the procedure, taking and changing, the latter is harder to carry out. Therefore, many teachers wanted to show this part visually, or let students have a hands-on experience of the fact that 1 ten is actually 10 ones (Ma, 1999).

Ma goes on to state:

> Scholars have noted that in order to promote mathematical understanding, it is necessary that teachers help to make connections between manipulatives and mathematical ideas explicit.

Concrete Instruction: Using Manipulatives

Manipulatives are great tools, but they have to be applied appropriately, otherwise, the students will not get the most value out of them. The key to filling gaps in basic math skills is to use manipulatives to teach the math, instead of using them only to perform operations. One of the best ways to introduce manipulatives to students is to teach the base ten number system and how numbers relate to each other. Depending on the level of your students' mathematical functioning, it may be necessary to start with something proportional, like base ten blocks. The place value mat and chips are more conceptual, because they are all the same shape and size.

Be sure your students are ready to use the place value mat and chips before you start using them. You will recognize students are ready to transition from the concrete to pictorial/representational (no longer proportional) when they no longer need the base ten blocks to answer questions such as, "What number, when added to 3, makes 10?" The following video activity demostrates an example of a pictorial activity.

Video Activity: Transitioning from Concrete to Representational to Abstract

Sample Activity: Subtraction with a Place Value Mat and Chips

This is an activity I used with a student who was having trouble subtracting multi-digit numbers. I sat down with him and pulled out my place value mat and chips. We looked over all of the pieces, and I asked him if he knew how to show the number 3 using chips. Right away, he figured out he should count out three of the 1-chips. I asked him where he thought he should put them on the mat. Again, quickly, he put the three chips in the ones column.

It looked something like this:

Hundreds	Tens	Ones
		① ① ①

Me: How can you take two away from that?

Student: *(He quickly took two of the chips.)* There's only one left!

Me: Good job! Now what do you know about the number 3?

Student: If you take 2 away from 3, you will only have 1 left

Me: Exactly! So tell me something else you now know about the number 3.

Student: 3 is 2 plus 1.

Hundreds	Tens	Ones
	10	1 1

Me: Tell me what number you see on the mat now.

Student: It's 12.

Me: How do you know it is 12?

Student: Because 10 plus 2 is 12. Also, the green dot says 10.

Me: *(It seemed like he had a good grasp of things so far.)* You're very observant. What happens when you try to subtract 3 from 12?

Student: That's where I get thrown.

Me: How should you start?

Student: First I have to draw a line through the 10 and make it a 9.

Me: How does drawing a line through a 10 make it a 9?

Student: I'm not sure, but that's what I have to do.

Me: What if we did something a little different? What if we look at the 10 differently? What do we know about tens in general?

Student: *(He looked at me as if I had grown three heads.)* I don't know.

Me: Well, let's look at the mat. What do you see on it?

Student: I see 12.

Me: And how do you know it is 12?

Student: Because it is 10 plus 2.

Me: But how do you know that is 12?

Student: Because! *(He was getting a little frustrated with me.)* There is 1 ten and 2 ones and that is 12.

Me: How do you know? What is it about the green disk that tells you it is 10? *(New math is all about the questions.)*

Student: Look. *(He began to gather more 1-chips.)* I'm going to count these out for you so you can see what I'm talking about.

And he did, he started counting them one at a time for me. He got to 9, and he kept right on going to 12.

Student: See? See, this is 12.

Me: But that is not what it looks like on the mat.

Student: That's because when we get to 10, we trade it in for a ... *(He paused. He looked at me, and he looked back at the mat, and then back at me.)* I get it!

Me: Show me.

Student: See, we don't have enough ones to subtract 3 of them. We only have 2, so we have to make more ones, but we can't just make them up out of nowhere. So instead of drawing a line through the 10 and writing a 9 above it, and adding it to the ones, we need to trade the green ten-chip in for 10 one-chips, and put all of them in the ones column.

Me: Are you sure?

Student: Yes. *(Then he put all 12 one-chips into the ones column.)* Now we have enough to subtract 3 ones. And that leaves us with 9 ones, so 12 minus 3 is 9.

Me: Nice job.

Student: You knew that all along, didn't you?

Me: I totally knew it. Now show me another one!

Use manipulatives to give students a firm foundation in number sense— and the relationships numbers have with each other. Provide students the opportunity to see that 1 ten is the same amount and has the same value as 10 ones; and that 3 tens and 7 ones has the same value as 1 ten and 27 ones; and that 6 tens and 9 ones has the same value as 5 tens and 19 ones, etc. Using concrete—physical—representations of the numbers gives students not only the conceptual understanding of the base ten system, but also allows them to see how decomposing and regrouping numbers can help them to visualize and manipulate numbers in ways that help them to solve problems.

Representational Instruction: Using Pictures

Once students have used manipulatives and appear to have mastered the concepts of the standards in the lessons, it is time to shift from the concrete (using manipulatives) to the representational (using pictures). One way to make sure students understand the math is to listen to them talk about the concepts they are learning. Conversing with other students encourages interaction among peers, and helps students develop their skills of explaining or justifying their problem-solving process. If students can't explain what they're doing, they probably don't understand it. Drawing a picture of what's happening is another way to get students to talk about what they're doing. Take the following problem, presented by the Adult Numeracy Network, for example:

Sample Activity: The Pencil Problem

In the morning, Ms. Wilkins put some pencils in a pencil box. After a while, she found that 1/2 of the pencils were gone. A little while later, she found that another 1/3 of the remaining pencils were gone. Still later, she found that another 1/4 of the pencils were gone. At that point, there were 15 pencils in the box. No pencils were ever added to the pencil box. How many pencils did Ms. Wilkins put in the pencil box in the morning?

A lot of students (and teachers) will attempt to find some kind of an equation, or algebraic expression to solve the problem. Here are a couple of example solutions:

Most common attempted solution:

$\frac{1}{2}p \times \frac{1}{3}p \times \frac{1}{4}p = 15$ Where p is the number of pencils in the box in the morning.

$\frac{1}{24}p = 15$ Multiply both sides by 24.

$p = 360$ Unfortunately, this solution isn't correct!

If you re-read the problem, you will notice that you are being told how many pencils are *gone* from the original amount. When the last group of pencils has been taken, you are told how many pencils *remain*. Decoding the problem is the first step to solving it. Because you are told how many pencils are left, you have to structure your solution from that perspective as well, by using the complement of each fractional part:

$\frac{1}{2}p \times \frac{2}{3}p \times \frac{3}{4}p = 15$ Where p is the number of beginning pencils.

$\frac{6}{24}p = 15$ Multiply both sides by 24 (for students with good number sense, you can reduce $\frac{6}{24}$ before multiplying).

$6p = 360$ Divide both sides by 6.

$p = 60$ This is the correct solution.

There is an even simpler method of solving this problem, and it relies on modeling:

Step 1: Draw the pencil box.

Step 2: Divide it in half, and color in one of the halves.

Step 3: Divide the remaining half into thirds, and color in one-third.

Step 4: Divide the remaining two-thirds into fourths, and color in one-fourth.

Step 5: Distribute the remaining 15 pencils to the 3 empty boxes. Fill in appropriate numbers in the rest of the boxes, and calculate the totals.

10	10	10
10	5	5
	5	5

Answer: The total number of pencils in the morning was 60.

Students who are visual or kinesthetic learners will benefit from the use of visuals and manipulatives when given a problem to solve. And for some students, just seeing more than one approach to solving a problem will help them to understand. The lesson here is twofold:

1. If it has not already been done during the student intake process, assess students' learning style preferences. Helping students discover their learning preference(s) can be one of the most validating experiences for them. It may help them to know that the reason they didn't catch onto these concepts before is that they might learn better when using a particular strategy, such as manipulatives or drawings. Refer to Appendix G to read more about learning styles and preferences.

2. You should practice differentiating instruction. You can make a student's learning challenges seem more manageable by simply presenting instruction in a different way or by allowing a student to access instruction and concepts in a way that makes them easier to comprehend. Learn multiple ways to present concepts and practice them with your students. Then it will get easier for you to try a different approach when one of your students isn't catching on.

Abstract Instruction: Using Numbers and Symbols

Once your students have demonstrated command of the concepts you have been teaching and they no longer have to rely on manipulatives or other representational drawings they are ready to transition to more abstract math. The following video activity demonstrates the use of abstract instruction to build on students' knowledge of fractions.

Use the fraction packets to demonstrate properties of fractions, along with how to begin performing operations with them. Students should understand the ways to use fractions to solve problems and should be able to use their understanding of fractions, proportional relationships, and operations in algebra to solve more sophisticated, abstract problems.

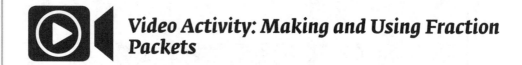

Video Activity: *Making and Using Fraction Packets*

The Math Teacher as a Facilitator of Learning

Students often say, "I don't need to know how to do this. I just want to get a job." Or, "I just want to go to technical school." But it should be part of our job as math instructors to help students see the relevance in what they are learning. Teachers need to present a case for learning math and explain how math fits into students' lives. According to researchers, adult education students will encounter real-world math in these four contexts:

- **Family or personal** is related to the student's role as parent, head of household, or family member. The demands include consumer and personal finance, household management, family and personal health care, and personal interests and hobbies.

- **Workplace** deals with the ability to perform tasks on the job and to adapt to new employment demands.

- **Community** includes issues around citizenship, and other issues concerning the society as a whole, such as the environment, crime, or politics.

- **Further learning** is connected to the knowledge needed to pursue further education and training, or to understand other academic subjects. (Manly, et al., 2006).

How do you convey the importance of math to your students? Ask students what math means to them and why they think they do or do not need to learn math. Talk about the math they encounter day to day. Connecting math concepts to students' everyday lives, makes it seem more relevant. When students solve page after page of addition, subtraction, multiplication, or division problems, they don't see any connection to real life. They need to see the real-world connections and have opportunities to practice and explore the math meaningfully.

How can learning fractions help with meal planning and budgeting? What kind of math will a student need to master to transition to the healthcare field? On any given day there is something math-related in the news: political polls, changing interest rates, sports scores, etc. How can you bring math and current events together in the classroom to help students better understand both? Try assigning activities that not only involve mathematics, but also a great deal of critical thinking, such as having each student research a news topic of importance to him, and relate it to math. This could involve averaging temperatures for a location, converting baseball statistics from percents to decimals to fractions, or even graphing gas prices to make predictions. Activities like these can become interdisciplinary—crossing over to language arts and/or social studies—and can provide a great deal of practical application. What other kinds of real-world math tasks can you prepare your students to face?

Teaching with Word Problems

One way to bring real issues into your math class is with word problems. Adapt word problems to fit contexts that are relevant to students and they are interested in. Word problems can be difficult for some students to understand, but they are on every high school equivalency test. It's important that students learn how to translate a word problem into an equation.

First demystify word problems by explaining that there are only three types:

1. Start unknown

2. Change unknown

3. Result unknown

In other words, to solve a word problem, first identify what is missing. The missing information is either a starting quantity, a type of change that occurs to that quantity, or the resulting quantity. Describing word problems in this way gives students a plan of attack. They know that when they encounter a problem, the first thing they need to do is read the problem and determine which type it is. Identifying the type of word problem tells the students what to solve for, and this can make a confusing word problem seem solvable. To familiarize students with this first step, have them practice solving simple word problems of each type.

If students need structure to help them follow the problem solving process, try using a graphic organizer. See Appendix E for an example.

 *Video Activity: Solving Relevant Word Problems*

Here is a simple example of each type of word problem:

START UNKNOWN

Trevor gave each of his three sons some money to spend on vacation. If each son got $30, how much total money did Trevor have to give them?

Because we do not know how much total money was given to the boys, <u>this is a start unknown problem</u>. To solve the problem, set it up like this:

$$x \div 3 = \$30$$

Let x = the total money Trevor gave to the 3 boys. To solve for how much total money Trevor had, we need to isolate the x on one side of the equation. Since we are dividing by 3 on the left, we do the inverse—or multiply both sides by 3. It may help your students to visualize the process by rewriting the problem like this:

$$\frac{x}{3} = \$30$$

Be careful not say something like "we need to multiply by 3 on each side so the 3s on the left cancel each other out." What is actually happening is that we are "reducing" the 3s. Look at it like this:

$$\frac{x}{1} \times \frac{3}{3} = \$30 \times 3$$

I have highlighted the original parts of the problem. Since any number divided by 1 is itself, and any number divided by itself is 1, we can simplify this problem. This will help students gain a conceptual understanding of the math.

$$x = \$90$$

The equation reduces to a simple multiplication problem. Now take the answer and insert it into the original problem to check the answer:

Does $\$90 \div 3 = \$30$? Yes, it does.

CHANGE UNKNOWN

Trevor gave his sons $90 to spend on vacation. He gave each son the same amount. If each son got $30, how many sons does Trevor have?

Because we do not know how many sons Trevor has, we do not know what to divide 90 by to get 30. So this is a change unknown problem. But let's set the problem up like before:

$$\$90 \div x = \$30$$

Look for the step that will isolate the x. In this case, it's helpful to rewrite this to demonstrate what is occurring:

$$\frac{\$90}{x} = \$30$$

To isolate the x, begin by cross-multiplying terms. Again, a rewrite may help students see what is actually being multiplied:

$$\frac{\$90}{x} = \frac{\$30}{1}$$

Cross-multiplying terms results in:

$$30x = \$90$$

Now divide both sides by 30 to isolate the x:

$$\frac{30x}{30} = \frac{\$90}{\$30}$$

Simplify:

$$x = 3$$

Trevor has 3 sons to give money to.

I chose cross-multiplication as the strategy to solve this problem because it led to a much simpler solution path. Consider the original problem:

$$\$90 \div x = \$30$$

If we isolate the x using the same strategy as the first solution, it would involve more steps. For example:

Multiply both sides by x:

$$\frac{90}{1} \times \frac{x}{x} = \$30 \times x$$

$$90 = 30x$$

$$\frac{90}{30} = x$$

$$3 = x$$

Result Unknown

Trevor gave his three sons $90 to spend on vacation. If each son received the same amount, how much spending money did each one get?

Because we do not know how much money each of the three boys received, this is a result unknown problem. This may be your students' favorite type of problem to solve.

$$\$90 \div 3 = x$$

Use the same strategy for solving this problem as in the other examples. This time, the x is already isolated.

$$\frac{\$90}{3} = x$$

Divide to simplify:

$$\$30 = x$$

This is a much easier process, which is why students prefer this type of problem.

Word problems are not going away. They are included in curricula at every grade level and on all the high school equivalency tests. Think of word problems as a good vehicle to demonstrate the usefulness of mathematics. Word problems based on real-life scenarios can help answer the question, "When am I ever going to use this?"

For example, look at the difference between these percent problems:

1. What is 25% of 73?

% × whole = part	percent formula
25% × 73 = x	substitute the given information into the formula
.25 × 73 = x	convert the percent to a decimal and then multiply the terms
18.25 = x	record your answer

2. You have two coupons for a department store, but you can only use one with your purchase. One coupon is good for $15 off your highest priced item, and the other coupon is good for 10% off your entire purchase. If you have a set of sheets for $55 dollars, 6 placemats at $4.99 each, and 4 Green Bay Packers freezer mugs at $19.99 each, which coupon is going to save you the most money?

Problem number 1 has no context. Students cannot relate it to their lives. But problem number 2 can be customized to appeal to your students. Which type of word problem is this? How would you model solving this word problem? A word problem like this can help you to demonstrate the process for decoding the problem and identifying the missing information.

This scenario offers an opportunity for you to help your students make some crucial connections. The first thing I do when tackling a word problem in class is to show it to my students and encourage them to begin working together—

collaborating—to decide on how best to begin solving the problem. This kick-starts students' critical thinking process. Give them some time to struggle with it, chew on it, and discuss it with each other. Do not shortchange students by interrupting this time with more questions. It's fine, and necessary, to be circulating and monitoring the discussions, but keep interventions minimal. As you notice what students have decided to do, ask questions such as, "What made you do that?" or "Will this method work every time?" or "Can you draw your solution on paper?"

Students might say, "I understand it, but I just can't explain it." If a student can't explain what she did to reach her solution, then she doesn't understand the problem or the solution. This is really important: Make sure students can explain and justify their work. Constantly repeat: "Show your work." What you really want is for students to show or tell you what they are thinking.

After they've spent some time looking at the problem, most students might recognize the need to list the price of each item, add the amounts together, and compare the totals. Some will do this before reducing the highest priced item by $15 for the first coupon—this is fine as long as they remember to take $15 off the total.

$15 off highest priced item	10% off entire purchase
$55 − $15 = $40.00	$55.00
$ 4.99	$ 4.99
$ 4.99	$ 4.99
$ 4.99	$ 4.99
$ 4.99	$ 4.99
$ 4.99	$ 4.99
$ 4.99	$ 4.99
$19.99	$19.99
$19.99	$19.99
$19.99	$19.99
$19.99	$19.99
Merchandise Total:	
	−10% =
TOTAL:	

Students who solve this problem quickly and correctly have great number sense—but this is rare. If none of your students are able to do this automatically, ask how they think the problem might be simplified. If no students venture a guess, encourage them to think about a method of simplifying it by asking a question such as, "Do you really want to add all those prices that end in $0.99?" Then prod them to think of strategies to reduce the prices, or to round the numbers to make "nice numbers," or numbers that are easier to work with.

If you don't like the term "nice numbers," have your students come up with a description they like. Explain that some numbers are easier to work with than others, such as numbers that end in 0 or 5.

Point out the groups of prices that end in $.99. Ask students if they know how to simplify them. Round the price up to $5 for each of the six $4.99 items. Ask students how much they will have to subtract from the total to make it accurate. ($0.01 × 6 items = $0.06) Round the price up to $20 for each of the four $19.99 items. Ask students how much they will have to subtract from the total to make it accurate. ($0.01 × 4 items = $0.04)

$15 off highest priced item		10% off entire purchase	
$40.00 $55 − $15 = $40.00	$ 4.99 $ 4.99 $ 4.99 $ 4.99 $ 4.99 $ 4.99	**$55.00** $55.00	$ 4.99 $ 4.99 $ 4.99 $ 4.99 $ 4.99 $ 4.99
6 × $5 = $30.00		**6 × $5 = $30.00**	
4 × $20 = $80.00 <u>$150.00</u> <u>− 0.10</u> **$149.90**	$19.99 $19.99 $19.99 $19.99	**4 × $20 = $80.00** <u>$165.00</u> <u>− 0.10</u> **$164.90**	$19.99 $19.99 $19.99 $19.99
Merchandise Total: $149.90		**$164.90**	
		−10% =	
TOTAL:			

Ask students if they know a simple way to calculate 10% of a number (move the decimal point one digit to the left). Or multiply by .10.
10% of $164.90 = 0.10 × 164.90 = $16.49

$15 off highest priced item	10% off entire purchase
$55 − $15 = $40.00	$55.00
$ 4.99	$ 4.99
$ 4.99	$ 4.99
$ 4.99	$ 4.99
$ 4.99	$ 4.99
$ 4.99	$ 4.99
$ 4.99	$ 4.99
$19.99	$19.99
$19.99	$19.99
$19.99	$19.99
$19.99	$19.99
Merchandise Total: $149.90	**$164.90**
	−10% of $164.90 = −$16.49
TOTAL: $149.90	**$148.41**

The best deal is to subtract 10% off the entire purchase price. The important lesson here is to understand how the math works so this process can be repeated in real situations. When students can set up the problem correctly and feel confident that they understand the math, then they can transfer that knowledge to other scenarios. This is conceptual understanding.

Talking about Math

The previous example points out how beneficial it can be for students to talk about solving a math problem before they are asked to find the answer. Adult students often learn best when they collaborate and learn from each other. This helps students develop not only critical thinking skills, but also critical communication skills. Speaking and listening to others, and giving and receiving feedback are essential skills for anyone, and they are vital in the adult education classroom.

"You can't subtract 9 from 7." "We're borrowing from the tens column." Phrases like these can drive teachers—and students—bonkers. It is no wonder students are confused after they hear phrases like these and then, during a lesson on integers, they see 9 subtracted from 7.

To avoid confusing students, think about what specific words to use when giving directions or instruction. Are you using a term that would make no sense if translated literally? Would it make sense to a student who is struggling with a processing or auditory deficit? Do your words mirror the examples the student is seeing? Are you using a term that you have used before in an entirely different way? Ask yourself how the student might interpret what you are saying. Choose your words carefully. One of the Standards of Mathematical Practice—in the CCRS—is "Attend to precision." This applies to teachers as well. Look at the example below of a student's work:

$$\text{Reduce } \frac{40}{72}.$$

$$\frac{40}{72} = \frac{40}{72} = \frac{40}{72} = \frac{40}{72} = \frac{40}{72} = \frac{40}{72}$$

Technically, the student did a fabulous job of reducing the fraction. And this is a great example of why we need to say exactly what we mean.

Here are a few strategies and activities designed to get students talking about the math they are doing. Create an atmosphere where it's okay for students to talk through their problem-solving steps, without criticism. Allow them to show their work and explain why they do what they do. Let the other students ask questions or comment. The more students talk about math, the more comfortable they will become in describing their own problem-solving processes. Here are some ideas to get students talking:

- Use a "Problem of the Day" or "Problem of the Week" to get students used to seeing different problems and talking about them, without the fear of failure. Solving word problems requires literacy skills as well as math skills. And when students discuss with their peers why they chose to work a problem the way they did, they not only practice using math terms, but they also demonstrate their understanding of the problem and its solution(s). When they receive feedback from their peers or when others show them how they have approached the problem differently, students can open their minds to seeing math in new ways.

- Use different types of problems, like puzzles or brainteasers, to pique their interest and show them that math can be fun. One way to get students engaged in talking about math is to present a challenge problem as they arrive. Make it challenging and fun—something students will look forward to doing. Not all of the work our students do has to be based in real life. Not to diminish the importance of connecting math to the real world, but sometimes, it's okay to toss them a brainteaser or puzzle. These types of problems can really get a good discussion going. It's great to hear the excitement in a student's voice when she explains to her peers how she solved a puzzle.

- Present students with a math standard and ask them to talk about it and come up with examples of it. Be prepared to define the math terms and language used in the standards.

Here is an example of how to use a standard. This standard is from the CCRS, Level D Functions:

> **8.F.3** Interpret the equation $y = mx + b$ as defining a linear function, whose graph is a straight line; give examples of functions that are not linear. For example, the function $A = s^2$ giving the area of a square as a function of its side length is not linear because its graph contains the points $(1, 1)$, $(2, 4)$ and $(3, 9)$, which are not on a straight line.

First, ask students to explain why the equation $y = mx + b$ will always define a linear function. Try this as a discussion starter at the beginning of the class period when transitioning from linear functions to quadratic functions. Students might describe or talk about how the slope (or rate of change) is constant.

- Consider making a math word wall. Jana Hazekamp, in *Why Before How*, provides a template for a math word wall. There are so many vocabulary terms in math that students may not understand. Often they are familiar with a term or have heard it before, but they may not know what it really means. A word wall helps to show students that math words and concepts are often connected. Ask students to contribute definitions and examples to the word wall so that the vocabulary becomes meaningful to them. If students are going to be encouraged to talk about what they're learning, you want to make sure they are using the appropriate terms. In conjunction with the word wall, you could also use math journals. Have students create a section in their interactive journals that is just for vocabulary. Here is an example of how to use a 3-column format:

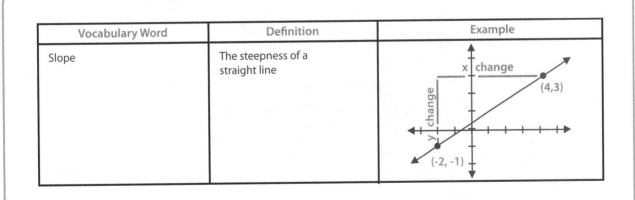

Vocabulary Word	Definition	Example
Slope	The steepness of a straight line	

What is Numeracy?

One of the most researched and widely respected published works on adult numeracy was produced by the National Center for the Study of Adult Learning and Literacy (NCSALL). The authors, Lynda Ginsburg, Myrna Manly, and Mary Jane Schmitt, released their work in December 2006, and 10 years later it still stands as the singular document responsible for the advancement and practice of adult numeracy. Ginsburg, Manly, and Schmitt state "numeracy is an array of mathematically related proficiencies . . . that include a connection to context, purpose, and use." Further, they state, "As quantitative and technical aspects of life become more important, adults need higher levels of numeracy to function effectively in their roles as workers, parents, and citizens."

There are many definitions of numeracy, but most are similar. For the purposes of this book, numeracy will be defined as the ability to work with and understand numbers and their relationships to each other.

The various definitions of numeracy have resulted in a variety of identified components of numeracy. Depending on the framework and perspective of the authors or researchers, they may identify the components of numeracy as specific skills to master, a grouping of similar topics or subjects, or a progression indicating which components need to be addressed first. Some components are highlighted below:

Ginsburg, Manly, and Schmitt (2006):
- Context: the use and purpose for which an adult takes on a task with mathematical demands
- Content: the mathematical knowledge that is necessary for the tasks confronted
- Cognitive & Affective: the processes that enable an individual to solve problems, and thereby, link the context and content

Other possible components of numeracy include the following:
- Flexibility
- Estimation
- Efficiency
- Awareness of relationships or patterns
- Determining reasonableness of results
- Prediction
- Mental computation
- Reflection
- Creativity
- Algorithms
- Speed

For discussion in this book, the components of numeracy will include the following:
- **Flexibility:** being able to understand a problem, recognize how to get the solution, and find a solution path at any point

- **Patterns:** seeing and manipulating patterns in order to solve problems

- **Representations:** using drawings and other tools to visualize mathematical concepts

- **Relationships:** knowing and using connections between numbers and concepts

- **Links:** applying past knowledge and/or solutions to new problems

- **Estimation:** being able to estimate a solution before computing it; checking the reasonableness of an answer

- **Mental math:** being able to perform computations mentally

- **Modeling:** using math to solve real-world problems

Flexibility

Clearly, there are a lot of different perspectives on the components of numeracy, but there are also many similarities. One reason to list flexibility first is that someone who is able to look at a problem in more than one way instead of depending on a formula is likely to have a greater sense of numeracy than someone who only knows how to use the formula. A flexible student is not thwarted by seeing a problem in a different form, like when the variable appears to the right of the equal sign. If a student freezes and doesn't know what to do next, there is no flexibility. Flexibility includes having a relaxed, "let's see where this goes" attitude. A flexible student is willing to listen to his peers describe how they solved a problem.

Sometimes students make links between a new problem and past problems or solutions. Also, give them the freedom to explore connections to their daily life. Students commonly see connections in geometric figures.

One student in my class realized a connection while we were working on circles on the coordinate plane.

I asked how much of the circle was contained in the first quadrant. One student replied, "one quarter."

Another student jumped up and yelled excitedly, "I get it! I get it!"

When I asked him what he got, he said, "I know now why when it's 3:15, they call it a 'quarter after.'"

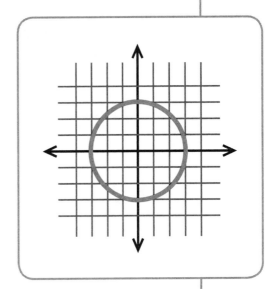

He was thrilled to be able to relate something we were studying in math class to something he had heard people saying, but never quite understood. It's those moments when we can see students developing a sense of numeracy.

Patterns

So much of math is about patterns, especially algebra. When students are able to detect a pattern and use it to describe other similar conditions, or use it to solve a problem, they demonstrate another crucial component of numeracy. One of the simplest examples of using patterns to solve problems involves linear functions. Consider the three function tables below:

Table 1	
x	**y**
−2	−1
−1	0
0	1
1	2
2	3

Function rule:
$f(x) = x + 1$

Table 2	
x	**y**
−2	6
−1	3
0	0
1	−3
2	−6

Function rule:
$f(x) = -3(x)$

Table 3	
x	**y**
−2	9
−1	7
0	5
1	3
2	1

Function rule:
$f(x) = -2x + 5$

Show students the tables. Ask them to identify the functions being applied to the *x* values in each of the three tables. The first two are relatively simple, the third may be a little trickier.

The grid on the next page shows another pattern. This is a representation of the coordinate plane. It looks a bit like a multiplication table, with factors across the top row and down the left column. The zeroes stand for the *x*- and *y*-axes. This grid demonstrates the answer to the question, "Why does a negative factor multiplied by a negative factor yield a positive product?" By looking at the 3rd quadrant of the coordinate plane, you see that just by following a pattern of ascending values in the left-half of the coordinate axes, the numbers increase as they proceed toward the bottom until, in the 3rd quadrant, there are only positive values.

x	−4	−3	−2	−1	0	1	2	3	4
4	−16	−12	−8	−4	0	4	8	12	16
3	−12	−9	−6	−3	0	3	6	9	12
2	−8	−6	−4	−2	0	2	4	6	8
1	−4	−3	−2	−1	0	1	2	3	4
0	0	0	0	0	0	0	0	0	0
−1	4	3	2	1	0	−1	−2	−3	−4
−2	8	6	4	2	0	−2	−4	−6	−8
−3	12	9	6	3	0	−3	−6	−9	−12
−4	16	12	8	4	0	−4	−8	−12	−16

Q2   Q1   Q3   Q4

Representations

This is another representation of the coordinate plane. In this grid, the factors are listed as ordered pairs (x, y). Now you see how you can use the coordinate plane to represent a variety of different equations, inequalities, and functions.

(−4, 4)	(−3, 4)	(−2, 4)	(−1, 4)	**(0, 4)**	(1, 4)	(2, 4)	(3, 4)	(4, 4)
(−4, 3)	(−3, 3)	(−2, 3)	(−1, 3)	**(0, 3)**	(1, 3)	(2, 3)	(3, 3)	(4, 3)
(−4, 2)	(−3, 2)	(−2, 2)	(−1, 2)	**(0, 2)**	(1, 2)	(2, 2)	(3, 2)	(4, 2)
(−4, 1)	(−3, 1)	(−2, 1)	(−1, 1)	**(0, 1)**	(1, 1)	(2, 1)	(3, 1)	(4, 1)
(−4, 0)	**(−3, 0)**	**(−2, 0)**	**(−1, 0)**	**(0, 0)**	**(1, 0)**	**(2, 0)**	**(3, 0)**	**(4, 0)**
(−4, −1)	(−3, −1)	(−2, −1)	(−1, −1)	**(0, −1)**	(1, −1)	(2, −1)	(3, −1)	(4, −1)
(−4, −2)	(−3, −2)	(−2, −2)	(−1, −2)	**(0, −2)**	(1, −2)	(2, −2)	(3, −2)	(4, −2)
(−4, −3)	(−3, −3)	(−2, −3)	(−1, −3)	**(0, −3)**	(1, −3)	(2, −3)	(3, −3)	(4, −3)
(−4, −4)	(−3, −4)	(−2, −4)	(−1, −4)	**(0, −4)**	(1, −4)	(2, −4)	(3, −4)	(4, −4)

Q2   Q1   Q3   Q4

This grid represents ordered pairs that could be plotted on the coordinate plane. These two representations of the coordinate plane can be used to generate discussions about patterns and representations. Think of some tasks you could ask your students to solve or discuss using one or both of these grids. Make a list:

TEACHING ADULTS: A MATH RESOURCE BOOK

1. Identify _____

2. Show _____

3. Describe _____

4. Predict _____

Later in this book you will have a chance to look at some examples of how functions can be represented on the coordinate plane. See chapter 10.

Relationships

Referring again to the Standards for Mathematical Practice, MP.1 is a good first practice for students to develop and use. It addresses the behaviors students need to use to solve every math problem.

MP.1: Make sense of problems and persevere in solving them.

The process of making sense of problems can look very different from one student to the next. One reason is that learning style differences may affect how students decode or understand problems. Manipulatives and representations can help some students to begin the process. As instructors, you need to help students find the entry point to a problem's solution. For example, if a student struggles with fractions, demonstrate the relationship between fractions and decimals using money as a concrete example. Another student may need to translate a word problem into an image or use manipulatives to represent the problem.

People learn in different ways, and they may even relate to numbers in various ways. One learner may demonstrate number sense in an entirely different way than another learner. A visual learner might see numbers in his head and be able to mentally break them apart and put them together again to make computations easier. See Appendix G for learning style inventories and resources.

Students may already know some methods to convert decimals to percents and vice versa. But it's important to understand not only how students have been taught to solve these conversion problems, but also what they understand about the relationship between decimals and percents. To figure a tip at a restaurant, for example, a person might first add or subtract a few cents to round the total to a whole dollar amount and then add whole dollars to the bill to approximate an appropriate amount. If students are used to guestimating percentages this way, they may have trouble accurately calculating percentages. Begin by teaching students to understand the relationship between decimals and percents, and then they can build from there.

These problem-solving structures show the relationship between decimals and percents three different ways:

$$\% \times \text{whole} = \text{part} \qquad \frac{\text{part}}{\text{base}} = \frac{\text{percent}}{100}$$

part $\Big/$ **p** — % in decimal form

whole $\Big/$ **w × %**

The first option presents the formula like a sentence; it is an equation that tells the student exactly what to do, regardless of what information is given. But this method might not make sense to all learners, so be prepared to present the formula in other ways, as well. The other two examples show the relationship between decimals and percents and how to convert from one to the other. A student may grasp one of these representations more easily than the others.

Links

Links speak to coherence—sequencing math instruction so students can see how conceptual understanding is built. The CCRS have identified three key shifts needed in math instruction for the 21st century:

- Focus
- Coherence
- Rigor

In an ideal situation, with a well-crafted curriculum, students will make the connections between what they have previously learned and practiced and the new math they are being asked to do. When students can see how mathematical concepts are related and how problem-solving strategies can be used to progress through a solution path, that is an indicator of developing numeracy skills.

Estimation

Estimation is a skill students need in order to decide whether a solution is reasonable. It's good practice for students to estimate what they think the answer will be or what range it might fall in. Start slowly if this is a skill students are not used to using. For example, if you are dividing a number in half, the quotient should be smaller than the dividend. Rounding numbers to a given place value can often help in estimating. Estimating may not be something students are going to use every day, but it has its purpose. If a student is studying for a high school equivalency (HSE) exam, he could use estimation to quickly narrow down the possible answers on a multiple choice question. Estimating can also help to build confidence. If a student concludes that an answer is close to what she estimated it would be, she can feel confident that she has calculated correctly.

Mental Math

Being able to decompose, recompose, and perform calculations mentally is a tremendous indicator of number sense or numeracy skills. When a student can visualize how a number or operation can be manipulated in order to make computations easier, she is able to work with fluidity and will not be limited to following the exact way the problem is presented. Students who master mental math are not easily frustrated and will likely be able to look at a problem from many different angles.

Modeling

When students can model, they are able to take an abstract mathematical concept and apply that conceptual framework to a relevant, real-world situation. Being able to find or create a model to describe a variety of situations—from data mining to geometry to algebraic reasoning—is proof that a student has a strong sense of numeracy and all its components. Modeling a math concept correctly shows complete understanding.

Teaching Number Sense

In *Building Powerful Numeracy in Middle and High School Students*, Pamela Harris says:

> Early intuitive number sense was not discovered until recently, so rather than building upon it, well-meaning teachers have often worked against it by emphasizing procedures and sacrificing the development of number relations. As a result, many learners abandoned their intuitive math sense in order to adopt the procedures being emphasized. They lose sight of how the numbers are related to each other (Harris, 2011).

The emphasis on procedural fluency has compromised overall math skills for decades. This reinforces how important it is to consciously and purposefully change the way math is taught. When you think about number sense, think about the process of cooking. At first, you might follow a recipe explicitly, but as you gain confidence and become more comfortable in the kitchen, you need to refer less often to the recipe. Number sense is a foundation for more complex mathematical competencies. With practice and a firm foundation in number sense, a student will be able to easily transition to more challenging and sophisticated tasks. And many foundational skills will become automatic, like a favorite recipe.

What if a student does not have intuitive number sense? Can numeracy and number sense be explicitly taught? Yes. What follows are some ideas to use to imbed explicit instruction for number sense development into your teaching.

Assessing Number Sense

First, get a quick idea of the strategies students know how to use when performing basic calculations. Start by asking them all to solve a problem that you have given them. While they work on this, walk around the room and observe their activity. Some students will look around at others and perhaps see them working, but they have no idea how to start. That's okay.

Try this problem: 47 + 19 = ?

Give students a minute or two to perform the calculation and then ask them for examples of how they attempted to solve it. Don't ask for the answer. This shows that you believe that the process is every bit as important as the content—or the answer. Ask for volunteers to share how they went about solving this problem.

You will likely have a student who looks at you and says, "I just added them." Time to dig deeper with your questioning. Ask the student to demonstrate or explain with a little more detail about the steps he took to reach a solution. Did he add 7 and 9 and carry the 1 to the tens position? This strategy applies a familiar algorithm, probably without any thought to why it works. But there's also a chance that the student was using another strategy. Until the student explains his thought process, it is equally likely that he could have been demonstrating some great number sense, or just applying an ancient algorithm.

Once the student has explained his process, open a discussion with the class to see if anyone used a different strategy to find a solution. You might get a response like this: "I took one away from 47 and added it to the 19 to make a nicer number to work with. So I changed the equation to 46 + 20." Brilliant! A student who thinks like this has had some exposure to number sense, or is able to rely on her innate sense of how to work with numbers. Not many students naturally think this way, but you can model and teach this process.

Ask if students used any other strategies to solve the problem. What you accomplish by expecting—and accepting—different problem solving strategies is to normalize the experience for those students who are aware they use a different thought process. You begin to build a safe space where their thoughts are respected and where they feel safe to discuss the processes they use. Have students talk in pairs or in groups, or have a student describe her problem-solving process for the entire class. This procedure has some valuable outcomes:

- Students learn their ideas are important, even if they do not reach the same solution as everyone else.

- Students begin to engage in the vital practice of giving and receiving feedback, without judgment.

- Students hear more of each others' voices, and less of yours.

Shifting Perspectives

There are several things you can do to correct or reteach topics that might be missing from an adult learner's conceptual understanding of mathematics. The shift from procedurally heavy instruction to emphasizing conceptual understanding of mathematics demands explicit instruction; it also means you can be creative in crafting assignments and problems.

For example, look at the difference between these two division problems:

Given what you know about division of whole numbers, find the missing digits:

$$8\overline{)1\square,5\square8}\quad\begin{array}{c}\square,5\square1\end{array}$$

vs.

Solve the following division problem:

$$8\overline{)12,568}$$

These example problems illustrate the change happening in math classes around the country. To successfully solve the first problem, students have to have an understanding of what is happening in the division process, as well as master decomposing and regrouping skills with numbers up to 100.

The first question draws on a student's command of number sense to take in the problem as a whole—before beginning to solve it. For example, the first thing to notice is the quotient terminates in 1 and has no remainder. The terminating quotient might mean a student can work backward to structure the solution.

Students may have no difficulty at all using the standard algorithm (long division), and that's great. But consider how much more they will learn if they first concentrate on the top problem, and describe all the steps they take to solve it.

Here's an idea of the process you might follow to talk through the problem with your class:

1. First, I see that the divisor is 8, the dividend ends in 8 and the last digit of the quotient is a 1. So I know the quotient does not have a remainder.

$$8\overline{)1\square,5\square8}\quad\begin{array}{c}\square,5\square1\end{array}$$

2. Looking again at the dividend and divisor, I see the first number (in the ten thousands place) in the dividend is a 1. I know we are only dealing with whole numbers, so the 2nd digit has to be between 0 and 9.

3. 8×1 gives me a single digit product, and 8×2 gives me a product with a tens digit of 1. 8×3 gives me a product with a tens digit of 2. So I think the first digit in the quotient is 1 or 2.

$$\begin{array}{r} 1,5\square1 \\ 8\overline{)12,5\square8} \end{array}$$

4. The second digit in the quotient is a 5, so I know that I have to have at least 4 left over when I subtract the first product. I'm going to guess that the first digit of the quotient is 1 and the second digit of the dividend is 2. This gives me 12 – 8, which leaves me with a remainder of 4.

$$\begin{array}{r} 1,5\square1 \\ 8\overline{)12,5\square8} \\ \underline{8} \\ 45 \end{array}$$

5. I drop down the 5, and know that 8×5 is 40, so 8 will go into 45 five times (confirmed by the 5 in the quotient).

$$\begin{array}{r} 45 \\ \underline{40} \\ 5 \end{array}$$

6. I multiply 8×5, and write that product under the 45. I subtract 40 from 45 and am left with 5.

7. I write my 5 below the subtraction. Then I need to decide what number is missing from the dividend.

$$\begin{array}{r} 1,571 \\ 8\overline{)12,568} \\ \underline{8} \\ 45 \\ \underline{40} \\ 56 \\ \underline{56} \\ 8 \\ \underline{8} \\ 0 \end{array}$$

8. This is where the end of the problem becomes the key to the solution. I need to put a number in the box that will leave a zero remainder (because I already know the dividend ends in 8 and the quotient ends in 1).

9. The only possible number to put in the box in the dividend is 6, because 8 goes into 56 evenly and there will be no remainder.

10. Lastly, I multiply the quotient by the divisor to check my answer.

11. The dividend is 12,568.

12. The quotient is 1,571.

One student told me he knew from the 5 in both the dividend and the quotient that he needed to put a number in the left-most box of the dividend that would result in a remainder of 4. He started in the middle, worked backward, and continued till he had the answer. There are many ways to approach a problem like this.

One of the most important lessons students can learn from an activity like this is that it is ok to start from the beginning, the end, the middle, or whatever part of the problem makes sense. As long as there is good reasoning behind the steps they take, they can be flexible.

Math Standards

Standards-Based Educational Reform

Most states with standards-based adult education programs have either adopted a set of adult education standards or they are in the process of adopting and/or aligning to standards. Some states have created or are in the process of creating their own standards. For example, Illinois uses a hybrid approach to combine the CCSS, CCRS, and its own content standards. To find the standards in your state—if they exist—you can go to your state Department of Education website.

The Office of Career, Technical, and Adult Education (OCTAE) has adopted the College and Career Readiness Standards (CCRS) to assist adult education programs with the shift to standards-based educational reform.

In order to determine what students should know by the time they graduate from high school, a great deal of attention was given to those foundational skills that should allow a student to transition successfully from secondary education to post-secondary education, or to the workplace (or both). The CCRS were written with the expectation that students who master them will be prepared for college or the workplace without the need for remediation. The CCRS are grouped into five levels, and the levels begin with the letters A–E rather than the six levels adult educators are familiar seeing with respect to the National Reporting System's (NRS) structure. The difference in the CCRS is slight, however, because the highest level, E, combines the top two levels of the NRS system into one.

Key Shifts

One of the most significant components of the CCRS is the emphasis placed on three key shifts:

1. **Focus**
   Generally speaking, instructors need both to narrow significantly and to

deepen the manner in which they teach mathematics, instead of racing to cover many topics. Focusing deeply on the major work of each level will allow students to secure the mathematical foundations, conceptual understanding, procedural skill and fluency, and ability to apply the math they have learned to solve all kinds of problems—inside and outside the math classroom (OCTAE, 2013).

2. **Coherence**

   Create coherent progressions in the content within and across levels, so that students can build new understanding onto previous foundations. That way, instructors can count on students having conceptual understanding of core content. Instead of each standard signaling a new concept or idea, standards at higher levels become extensions of previous learning. The focus on understanding numbers and their properties through the levels also exemplifies the progression from number to expressions and equations and then to algebraic thinking (OCTAE, 2013).

3. **Rigor**

   There needs to be equal measures of conceptual understanding of key concepts, procedural skill and fluency, and rigorous application of mathematics in real-world contexts. Students with a solid conceptual understanding see mathematics as more than just a set of procedures. They go beyond "how to get the answer" and can employ concepts from several perspectives. Students should be able to use appropriate concepts and procedures, even when not prompted, and in content areas outside of mathematics (OCTAE, 2013).

Incorporating these key shifts into instructional practices may seem overwhelming or cumbersome, but for students to succeed, this needs to be done.

Two of the shifts—focus and coherence—will be featured significantly in upcoming discussion and activities. Look at this example of the mapping of the lowest math standards, grade level expectation (GLE) K-1 or Beginning ABE Literacy:

The domains in this level consist of the following:

- Counting and Cardinality
- Operations and Algebraic Thinking
- Number and Operations in Base 10
- Measurement and Data
- Geometry

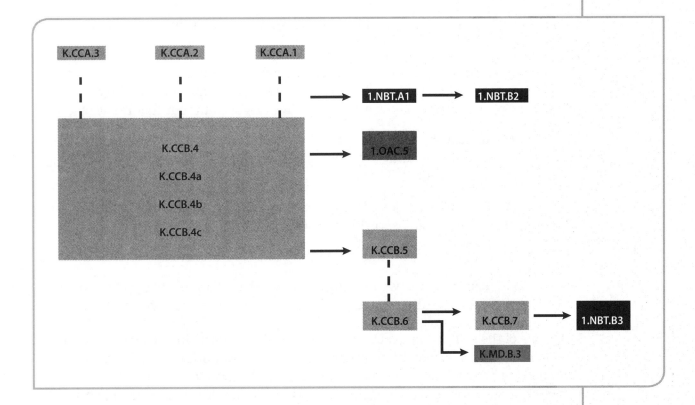

KEY:

K.CC.A.1: Count to 100 by ones and by tens.

> **1.NBT.A.1:** Count to 120, starting at any number less than 120. In this range, read and write numerals and represent a number of objects with a written numeral.

> **1.NBT.B.2:** Understand that the two digits of a two-digit number represent amounts of tens and ones.

K.CC.A.2: Count forward beginning from a given number within the known sequence (instead of having to begin at 1).

K.CC.B.3: Write numbers from 0 to 20. Represent a number of objects with a written numeral 0–20 (with 0 representing a count of no objects).

K.CC.B.4: Understand the relationship between numbers and quantities; connect counting to cardinality.

> **K.CC.B.4a:** When counting objects, say the number names in the standard order, pairing each object with one and only one number name and each number name with one and only one object.

> **K.CC.B.4b:** Understand that the last number name said tells the number of objects counted. The number of objects is the same regardless of their arrangement or the order in which they were counted.

> **K.CC.B.4c:** Understand that each successive number name refers to a quantity that is one larger.

1.OA.C.5: Relate counting to addition and subtraction (e.g., by counting on 2 to add 2).

K.CC.B.5: Count to answer "how many?" questions about as many as 20 things arranged in a line, a rectangular array, or a circle, or as many as 10 things in a scattered configuration; given a number from 1-20, count out that many objects.

K.CC.B.6: Identify whether the number of objects in one group is greater than, less than, or equal to the number of objects in another group, e.g., by using matching and counting strategies.

K.MD.B.3: Classify objects into given categories; count the numbers of objects in each category and sort the categories by count.

K.CC.B.7: Compare two numbers between 1 and 10 presented as written numerals.

1.NBT.B.3: Compare two two-digit numbers based on meanings of the tens and ones digits, recording the results of comparisons with the symbols >, =, and <.

Some of the skills targeted in this level and in these domains include counting, naming numbers, matching numbers to a specific quantity, and recognizing and using the symbols (numbers) in meaningful ways. Treating the domains as separate entities is no longer viewed as a best practice. The shifts in math instruction mean less emphasis is placed on procedures and more is placed on connections between related concepts. Mathematical understanding is emphasized over following steps in an algorithm—and a coherence map is one way to show that.

Consider the following coherence map for the domain of Counting and Cardinality (ABE Beginning Literacy level) as just one example. While the CCRS do not include this domain, many other standards do, so it is included here. There are two types of connectors in this coherence map: the green dashes signify that two standards are closely related and the black arrows indicate a student will have difficulty mastering the concepts in each successive standard without first mastering this content in the current standard.

Do you agree with the way standards are linked? Are there relationships or connections that are missing or need to be re-evaluated?

The benefit of a coherence map is that it forces you to consider not just the scope of what you will be teaching, but how the concepts connect. This helps thwart the "mile wide and inch deep" philosophy. Coherence is crucial to not only show students how mathematical concepts are related, but also to give them a sense of when and how to apply previous learning. Using coherence as a guide to structure your instruction will help students move past their reliance on formulas.

Looking at the coherence map, you might imagine how you could create lessons that contain the elements of the connected standards. Making a coherence map will help you to be mindful of the conceptual connections. If you are thinking about building your own coherence maps, check out the links and resources in Appendix G.

Math Levels

There are many different ways to describe the levels of math work. Three that are commonly used are Smith and Stein's Levels of Cognitive Demand, Webb's Depth of Knowledge, and Bloom's Taxonomy.

Levels of Cognitive Demand

When examining mathematical tasks, there are four levels of cognitive demand (Stein and Smith, 1998):

LOW-LEVEL
1. Memorization
2. Procedures without connections to concepts or meaning

HIGH-LEVEL
3. Procedures with connections to concepts and meaning
4. Doing mathematics (problem solving)

Webb's Depth of Knowledge

The GED® Test, the HiSET®, and the TASC test were developed using Webb's Depth of Knowledge (DOK). There are four levels of DOK. DOK refers to the level of cognitive demand required to answer a question correctly. About 20 percent of the questions on the GED test are DOK level 1. These types of questions require simple recall of knowledge. The 2014 TASC test included mostly level 1 and 2 questions, but it evolved to focus on more level 2 and 3 questions in 2016. The majority of questions on the GED, HiSET, and TASC tests are level 2 or 3; there are no level 4 questions on the tests.

The following chart describes the DOK levels and provides examples of question types at each level.

Depth of Knowledge
Level 1
Requires students to recall or observe facts, definitions, or terms. Involves simple one-step procedures. Involves computing simple algorithms (e.g., sum, quotient). Examples: • Recall or recognize a fact, term, or property • Represent in words, pictures, or symbols a math object or relationship • Perform routine procedure like measuring
Level 2
Requires students to make decisions of how to approach a problem. Requires students to compare, classify, organize, estimate, or order data. Typically involves two-step procedures. Examples: • Specify and explain relationships between facts, terms, properties, or operations • Select procedure according to criteria and perform it • Solve routine multiple-step problems
Level 3
Requires reasoning, planning, or use of evidence to solve problem or algorithm. May involve activity with more than one possible answer. Requires conjecture or restructuring of problems. Involves drawing conclusions from observations, citing evidence, and developing logical arguments for concepts. Uses concepts to solve non-routine problems. Examples: • Analyze similarities and differences between procedures • Formulate original problem given situation • Formulate mathematical model for complex situation
Level 4
Requires complex reasoning, planning, developing, and thinking. Typically requires extended time to complete problem, but time spent not on repetitive tasks. Requires students to make several connections and apply one approach among many to solve the problem. Involves complex restructuring of data, establishing and evaluating criteria to solve problems. Examples: • Apply mathematical model to illuminate a problem, situation • Conduct a project that specifies a problem, identifies solution paths, solves the problem, and reports results • Design a mathematical model to inform and solve a practical or abstract situation

(left column label: Mathematics)

The Mathematics table is used with permission of Dr. Norman L. Webb from the University of Wisconsin Center for Educational Research.

Bloom's Taxonomy

Bloom's Taxonomy is likely one of the most widely known classifications of educational topics. Bloom divided his taxonomy into three domains: cognitive, affective, and psychomotor. These domains are sometimes referred to as knowing/head (cognitive), feeling/heart (affective), and doing/hands (psychomotor). Bloom recognized that in order to make learning truly transformative, it should involve the whole person. As with Webb's DOK, Bloom's Taxonomy demonstrates how learning can be differentiated with respect to learning styles and learner preferences. Bloom's Taxonomy also emphasizes what students are able to *do* with their learning. This model expects students will use their learning, and will exhibit specific and observable behaviors while integrating instruction. While Bloom first published the taxonomy in 1956, a revised version, written by his former students, was published in 2000. The Revised Taxonomy now includes a final level of *doing* called "Creating."

Most people are accustomed to seeing Bloom's Revised Taxonomy as a pyramid, like in the image below left. But the image on the right provides a new perspective. The inverted pyramid puts more focus on the higher order thinking skills and less on the low-level skills of remembering, understanding, and applying. Bloom's Taxonomy is based on what students can do with what they learn. With "Creating" at the top of the new pyramid, students are expected to take what they have learned and use it to create something new.

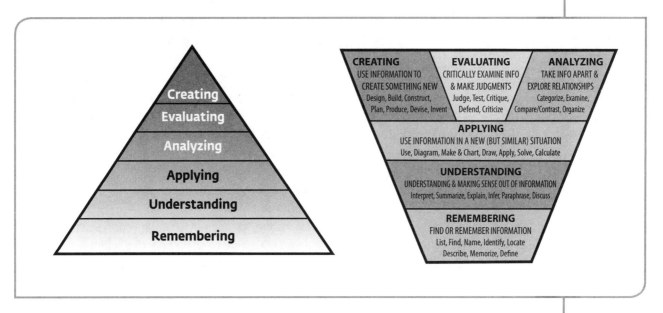

Reverse pyramid: Jessica Pilgreen, 2012

This chart shows how Bloom's Taxonomy applies specifically to math learning:

Bloom's Revised Taxonomy: Mathematics

<table>
<thead>
<tr><th></th><th>New Terms</th><th>Actions</th><th>Learning Activities</th></tr>
</thead>
<tbody>
<tr>
<td rowspan="3">Higher-order thinking</td>
<td>Creating

(Putting together ideas or elements to develop an original idea or engage in creative thinking).</td>
<td>Designing
Constructing
Planning
Producing
Inventing
Devising
Making</td>
<td>Creating: (Generating new ideas, products, or ways of viewing things) How could we determine the number of pennies in a jar without counting them? Apply and integrate several different strategies to solve a mathematical problem. Design a new monetary system or an experiment for establishing … Designing, constructing, planning, producing, inventing.
Invent a machine to do a specific task. Develop a menu for a new healthy foods restaurant.</td>
</tr>
<tr>
<td>Evaluating

(Judging the value of ideas, materials and methods by developing and applying standards and criteria).</td>
<td>Checking
Hypothesizing
Critiquing
Experimenting
Judging
Testing
Detecting
Monitoring</td>
<td>Evaluating: (Judging the value of a product for a given purpose, using definite criteria) Develop a proof … and justify each step …, Using a definition … determine …
Justifying a decision or course of action, checking, hypothesizing, critiquing, experimenting, judging
What criteria would you use to evaluate if your answer is correct? Prepare a list of criteria to judge… Evaluate expressions.</td>
</tr>
<tr>
<td>Analyzing

(Breaking information into its component elements to explore relationships).</td>
<td>Comparing
Organizing
Deconstructing
Attributing
Outlining
Structuring
Integrating</td>
<td>Analyzing: (Breaking information into parts to explore understandings and relationships) Given a math word problem, determine the strategies that would be necessary to solve it. Write a paragraph describing the relationship …, How does … compare to …
Comparing, organizing, deconstructing, interrogating, finding
Design a survey to find out … Graph your results. Use a Venn Diagram to show how two topics are the same and different. Translate between visual representations, sentences, and symbolic notation. Make predictions based on experimental or statistical data.</td>
</tr>
</tbody>
</table>

	New Terms	Actions	Learning Activities
Lower-order thinking	**Applying** (Using strategies, concepts, principles and theories in new situations).	Implementing Carrying out Using Executing	***Applying:*** (Using information in concrete situations) *Compute the area of actual circles. Use the given graph to ..., Choose and describe the best method to ...* Using information in another familiar situation, implementing, carrying out, using, executing *Draw a diagram which shows these fractions or take photographs of the fractions. Determine measures of central tendency and dispersion. Write a journal entry. Write an explanation about this topic for others.*
	Understanding (Understanding of given information).	Interpreting Exemplifying Summarizing Inferring Paraphrasing Classifying Comparing Explaining	***Understanding:*** (Grasping the meaning of material) *Given the mathematical formula for the area of a circle, paraphrase it using your own words. Select the graph that illustrates* Explaining ideas or concepts Interpreting, summarizing, paraphrasing, classifying, explaining *Find items that you can use to show the fractions. Retell or write in your own words... Report to the class... Write a summary report of the event.*
	Remembering (Recall or recognition of specific information).	Recognizing Listing Describing Identifying Retrieving Naming Locating Finding	***Remembering:*** (Remembering previously learned material) *State the formula for the area of a circle. State the rule of ..., Explain and use the procedure for ...* Recalling information, recognizing, listing, describing, retrieving, naming, finding, locating *List the fractions you know and can show. List the attributes of your shape. Make a concept map of the topic. Make a chart showing ...*

Credit: Milwaukee Mathematics Partnership, Center for Mathematics and Science Education Research, University of Wisconsin Milwaukee

When we look at a chart like this, it is easy to see how important coherence is; the things we look for—what students can do with the math they are doing—build on and flow from lower level thinking and activities to the higher level.

To embrace new instructional strategies and provide opportunities for students to develop higher order thinking skills, we will likely have to change the way we prepare for classes. I used to prepare by examining the section of the text I was using to develop an engaging and informative lecture that included opportunities to demonstrate skills for the students. Now that model seems very ineffective. Sure, some of the students grasped what I was teaching, but many didn't.

One afternoon while covering a study hall, the Driver's Ed teacher came in and took two students with him to practice behind-the-wheel driving. A lightbulb went off! I realized how unrealistic I was being to think my students would be able to learn math just from hearing me talk about it and watching me do it. They certainly couldn't learn to drive a car that way. Now I understand that I need to give students a chance to get their hands on the math and do it themselves.

The less time an instructor spends lecturing, the more time students have to work. There is a time and a place for explicit instruction, but I use it mainly to teach problem solving. Students can build a strong conceptual foundation while doing group activities. The latest research about math instruction emphasizes the importance of "number talks." Engaging adult students in discussion— in small groups or as a class—can help them to comprehend even the most complex concepts.

Math talk also relates to Bloom's levels 1–5, and perhaps, even level 6. It's a model that transitions to college and careers as well. Use number talks to observe, assess, redirect, and create engagement. This guide for crafting questions according to Bloom's Taxonomy can help you to create appropriate topics for your students to discuss:

Questions at the lower levels are appropriate for:

- evaluating students' preparation and comprehension
- diagnosing students' strengths and weaknesses
- reviewing and/or summarizing content

Questions at higher levels of the taxonomy are usually most appropriate for:

- encouraging students to think more deeply and critically
- problem solving, encouraging discussions
- stimulating students to seek information on their own

This table illustrates the flow from lower to higher level tasks and activities—as they relate to fractions and 3D shapes. This is also a good example of how an instructor can differentiate instruction or create varying expectations for students learning at a different pace.

Fractions with Bloom's Revised Taxonomy	3D shapes with Bloom's Revised Taxonomy
Remembering List the fractions you know and can show.	**Remembering** List the attributes of your shape.
Understanding Find items that you can use to show the fractions.	**Understanding** Find items that you can use to show the shape.
Applying Draw a diagram which shows these fractions or take photographs of the fractions.	**Applying** Draw a diagram or take a photograph of the shape.
Analyzing Design a survey to find out which fractions are easy and which are hard. Graph your results.	**Analyzing** Identify where the shape is found in the classroom and school.
Evaluating Choose a diagram or picture to represent the hardest fractions to use in a game.	**Evaluating** Tell why your shape is used in the places it is.
Creating Create a power point presentation game for others to play.	**Creating** Create an item that includes all or part of your shape. Draw and label your design.

Credit: Milwaukee Mathematics Partnership, Center for Mathematics and Science Education Research, University of Wisconsin Milwaukee

6

Teaching Adults Math

The Reluctant Learner

Mark H. Ashcraft defines math anxiety as "a feeling of tension, apprehension, or fear that interferes with math performance" (2002).

Some research has been done on the topic of math anxiety, and there is evidence to suggest a connection between early experiences in math and future math performance—or at least fears related to future math performance.

Of note to adult education teachers: "knowledge that a teacher's math anxiety can affect his/her students' math achievement suggests that we also need to ensure teachers feel fully confident in their preparation to teach math" (Beilock and Willingham, 2014).

Overcoming Math Anxiety

How do you feel about heights? Snakes? Bridges? Spiders? Most of us know that fears are not completely rational—many people who fear heights have never had a bad experience on a ladder or in a tall building. But the sweaty palms and rapid heart rate are very real.

You can probably recall someone admitting, "I am not a numbers person" or "I have always been bad at math," as if being bad at math is socially accepted. We have given (and received) the message that some people just don't have the ability to be successful at math. Students may use that as an excuse to not try. However, some people suffer from true math anxiety.

A two-pronged approach is needed to overcome math anxiety: First acknowledge it and then address it. When students express (verbally or nonverbally) a strong dislike of math, or a history of failing at math, they need more than a pat on the shoulder and a dismissive response. If a student shows signs of math anxiety, ask questions to find out exactly what he is experiencing:

- What happens to you physically when you attempt to do math?

- What happens to you emotionally when you attempt to do math?

- How do you feel when you have to demonstrate or describe your understanding of a problem?

- How do you feel when you take a math test or any test?

It makes sense that a student with math anxiety, depending on the extent and scale, may not perform as well as a peer without math anxiety. However, it is important to keep in mind that even a very bright student can be affected by math anxiety. Resiliency and coping skills can affect how much a student's performance suffers. Many adult education students have a history of being embarrassed or ridiculed by their peers—or worse, by a teacher. They may subconsciously associate school, teachers, or math class with negative feelings based on bad experiences in their past. Some students may have undiagnosed learning or processing disabilities, or some may have struggled with mathematical concepts from their earliest exposure.

Beilock and Willingham say that in the United States, about "25 percent of four-year college students and up to 80% of community college students suffer from a moderate to high degree of math anxiety" (2014). Overall, researchers estimate that half the U.S. adult population suffers from math anxiety. If that's true, how many of your students are having trouble managing their stress?

Human beings have three innate reactions to stress: Fight, flight, or freeze. It is unlikely students will lash out physically when given a math activity or assignment, but the other two reactions are quite common. Some students may walk out or give up. Other students may just freeze up and be unable to even take in what is being said.

Think of a time you experienced something that triggered your fight, flight, or freeze response. How did you feel physically and emotionally during the experience? How long did it take for the tension to subside? How were you able to resolve the situation that precipitated the stress reaction? How many of your students possess exceptional impulse control or are capable of working through such a difficult moment without assistance? Everyone has different ways of responding to anxiety and fear.

The key to confronting and neutralizing anxiety is to first identify the source, and then work to find productive solutions. Choose your words carefully; sometimes the use of negative language—especially from an authority figure—is like adding gasoline to a smoldering fire.

When a student begins to experience the first signs of anxiety, the response from those around him often determines whether the anxiety will escalate or subside. As soon as you notice, acknowledge that you see the student getting upset, and say something like, "You look like you're upset about something. Is there anything I can do to help?" Let the student know you have noticed his struggle, but also that this could really be a different kind of learning experience.

- **Fight:** To be clear, in this context, a fight response probably will not include a physical threat or altercation. If that is a concern in your classroom, speak to your administrator, and respond as necessary to keep yourself and your students safe. The fight response to math anxiety is more of an internal impulse to resist what is happening. Someone experiencing this response wants to stop feeling the way she is feeling, and looks for an immediate way to resist what is happening. The student may have stopped listening or she may be doodling or texting to avoid paying attention.

  You might not see exactly what the student is going through, but it could be obvious that she has "checked out," and any attempts to re-engage her are not successful. In this situation, attempt to connect with the student— talk to her about what you observed, and discuss what she is experiencing. It may help to remind your students that, no matter what, math is not going to hurt.

  If you let students know that your classroom environment is a safe place where risks can be taken and everyone is supported, you might not see a student react this way. The most helpful response to someone who appears to be experiencing anxiety is to offer support. Support can take many different forms. It might be walking among your students, offering verbal words of encouragement, or it might be making eye contact with a student to acknowledge the signs of struggle. Saying something like, "I think some of you are still struggling with this—who can come up with another situation we can use to apply this lesson?" both normalizes what your student is feeling, and lets him know others are struggling as well.

- **Flight:** The flight response could actually mean that a student gets up and leaves the classroom (depending on your class rules, you may or may not respond to this). Other flight responses might be using humor to distract you or others from the lesson. You may notice a student avoiding the work by staring out a window or down at her feet. If you notice increasing levels of anxiety among your students, this could be a good time to take a break and check their understanding. Have students talk in pairs or in groups about what you have been teaching. "Tell your partner what you just heard," is a good way to get conversations started.

- **Freeze:** The third, and more obscure response to anxiety is to become paralyzed. Have you ever had an experience where you were so overcome with stress that you could not decide what to do next? Some students, especially those who may have had negative classroom experiences in the past, may react to their math fears by freezing up. Reassure students regularly that the classroom is a safe place where they can ask any question and say whatever is on their minds. Be careful not to chastise students for giving an incorrect answer, instead focus on the process. It's sometimes easier to have another student demonstrate the correct process. That's less likely to make a student feel like you are the enemy or remind the student of a bad experience in his past.

The good news is that you have control over some of the main causes for math anxiety. According to Marilyn Curtain-Phillips (1999), "imposed authority, public exposure, and time deadlines" produce an inordinate amount of anxiety in students. If you act more like a learning partner than an authority figure, you may be more approachable and students may be more likely to admit their struggles. Likewise, instead of having students demonstrate problems in front of the whole class, have them work in pairs or small groups until they are comfortable with the problem solving process. Procedural fluency is a target, but first students need to gain confidence by mastering the content.

Though high school equivalency exams are timed, timing is unnecessary pressure when students are first trying to learn important skills and concepts. They can prepare for timed tests later. With practice, students can learn to do the math more quickly.

According to Beilock and Willingham (2014), "enhancing basic numerical and spatial processing may help guard against the development of math anxiety." What exactly does this mean? Learners with good number sense are less likely to suffer math stress. Being able to recognize and understand numbers, place value, number relationships, etc., will reduce the likelihood a student will develop math anxiety. If a student comes to you with math anxiety, devote some time to building his number sense. This will improve his confidence and reduce his anxiety.

When Will I Ever Use This?

Instruction—especially math instruction—has to be relevant. Outside of class or a high school equivalency exam, a student will probably never need to graph the solution to a quadratic equation. But, it is almost certain that, regardless of their chosen paths, your students will be required to problem solve, use critical thinking, and apply what they know to new situations. So, when your students ask you, "When will I ever use this?" tell them that they are learning skills and processes that they will use in the future.

Math class is not just about math skills. It's also about

- How to be patient when you're unsure how best to proceed
- How to evaluate if a solution to a problem is reasonable
- How to communicate with others in a professional or academic setting
- How to begin to solve a problem when you have no idea how it will turn out

You are also modeling persistence, which happens to be one of the Standards for Mathematical Practice.

The Reluctant Math Teacher

The shift to more rigorous math content has been difficult for many instructors. Adult education programs must often rely on volunteer tutors, or instructors who have had no formal math teacher training. But keep in mind, you are not alone. Every adult education program and many instructors are facing the same challenges: mastering new content and instructional strategies. Following are some suggestions that may help you to feel more comfortable about teaching math.

First, you might consider developing a Professional Learning Network (PLN) of some colleagues who specialize in math. If you have attended a really useful professional development session at a conference, email the presenter. Look for an experienced colleague who can become a mentor. Take a class. Watch YouTube. Follow teachers on social media. Look for blogs written by other math instructors. The internet has almost endless resources for professional development. See the list on page 65 and Appendix G for some online resources.

I Do Not Have a Degree in Math

Before you can expect to relieve students' anxiety, you must relieve your own. There are things you can do to become more at ease with math—even if you are not a math teacher or you have to teach unfamiliar concepts. Research documents and demonstrations on how to teach math content that is based on rigorous standards. View YouTube videos that demonstrate math concepts you feel unsure about. The better prepared you are, the less anxiety you experience, and the less chance there is you'll pass that anxiety on to your students.

If you are not comfortable teaching math, then work on building your number sense. This will help you the same way it will help your students. The more comfortable you are with the content you teach, the less stress you will pass on to students.

How Can I Teach What I Do Not Know?

Some teachers feel anxiety about looking foolish in front of their classes. With the new, more rigorous standards, more and more teachers are saying, "How am I supposed to teach math that I do not know how to do?" It's good to be honest about your feelings, but what are you really afraid of? Are you worried about yourself or about whether you can help your students to pass a test, get into college, or get a good job? If you are more worried about your students, use your fear and anxiety as a frame of reference to focus on the math most likely to help them succeed. Think about what you say to your students when they are afraid they may not understand complex concepts. As you transition from concrete to representational to abstract math, the work does get more rigorous and complex. How do you feel when you try something scary and new? What helps you reduce anxiety in new situations?

If you don't know where to start, try these tips:

- There are an almost infinite number of resources on the web, and more are being developed every day. These are some of my go-to sites:

  ▷ **Lesson Planet** (www.lessonplanet.com)
    This site has searchable lessons (by grade level or standard) by state, teacher reviewed resources, and a community of teachers ready to give advice and support.

  ▷ **Greg Tang Math** (www.gregtangmath.com)
    Try the game: Kakooma. You need an internet connection to play this game in class with your students. The game is based on number sense—the ability to rapidly and accurately compose and decompose numbers. Playing the game can help you—and your students—feel more comfortable working with numbers.

  ▷ **Purple Math** (www.purplemath.com)
    This site has courses covering math topics geared toward learners of all levels K-16. The site also has links to other reviewed sites that offer math content and instruction.

  ▷ **Illustrative Math** (www.illustrativemathematics.org)
    This site can be searched using any math topic or standard number. It also has tabs dedicated to progressions to help you structure your lessons with respect to coherence. The professional learning section offers professional development resources.

  ▷ **YouTube** (www.youtube.com)
    There are many professional development video courses available on YouTube. Just search for the topic you are interested in.

  ▷ **Virtual Manipulatives** (bit.ly/1M54wZg)—Use the virtual algebra tiles for front-of-the-room instruction, or instead of class sets. This site also has virtual base 10 blocks, number lines, and other tools.

  ▷ **COABE Repository** (www.coabe.org/educatorresources)
    COABE conference presentations and materials are uploaded here, and members can access them for free. Search the Adult Educator's Repository by content area.

  ▷ **LINCS** (community.lincs.ed.gov)
    The Literacy Information and Communication System (LINCS) site hosts several discussion groups—including Math/Numeracy. This is a great place to connect with peers and ask or answer questions about teaching math to adults.

  ▷ **ProLiteracy EdNet** (www.proliteracyednet.org)
    This site offers techniques, activities, and online professional development courses for adult education instructors.

- Talk with peers. Just like math talks help learners, talking about teaching math with other teachers will help you. Ask peers how they teach the topics you struggle with.

- Be responsible for your own professional development. Don't wait until the end of the year to get in your professional development hours. There are many resources available in print and online. Talk to your administrators to see if you can get professional development credit for reading a book, taking on online course, or viewing instructional videos.

- Consider joining the Adult Numeracy Network (ANN). They publish a newsletter for members, and they offer a full-day preconference session at the annual COABE conference. ANN also sponsors the Practitioner Research Project which seeks to engage instructors in shaping best practices by conducting research within and among their classes. For more information on ANN, see Appendix G.

- There are more resources listed in Appendix G.

My Students Know More Than I Do

Some students might know more than you about certain topics in your instruction. That's not necessarily a bad thing. It could be empowering to flip the classroom and have your students share their experience and knowledge with you and their peers.

In the Chinese language, there is no symbol for crisis. Crisis is represented by the symbols for the two sides of a crisis: danger and opportunity. Think of it this way: if a student catches a mistake in your instruction or asks a question that you have no idea how to answer it can feel scary, but it's also an opportunity. Use the opportunity to share the learning experience with your students.

We ask our students to take risks and to trust that their experiences in your class will be profoundly different from their previous experiences. Trust is a two-way street. Make your students partners in learning with you. If you are able to admit to your students, "I have always had trouble with trigonometry," but you show a willing attitude to tackle it, you become a model for positive thinking.

Teaching Tools

Math Journals

Writing math journals is a great way for students to translate your instruction into their own words and/or diagrams. You can also use them as daily, weekly, or monthly assessments.

There are many different strategies to develop math journal activities. Some teachers have students use one throughout their time in a class, other journals are subject based, and others are unit or lesson based. The more opportunities students have to get into their journals—whether to add new content, or search for existing content—the better they retain the information contained in them. Most teachers who use math journals use templates for the majority of their content entries—this is a good idea, as long as you still provide an opportunity for students to translate the instruction into their own words. When you review students' journals, you will get a clear picture of what they understand and what they still need help with.

You may choose to grade your students' math journals. If you do, be careful. You do not want to dissuade your students from writing about math in their own words. How you describe the project of making and keeping a math journal is crucial to how your students will feel about doing it. If you are going to grade the journals, tell students how you will grade them and provide them with the rubric you will use. If you don't grade math journals, you can still have students turn them in on a weekly basis so you can check on how they are doing and give feedback. This provides a way to have private conversations with students about the math, so that they don't feel embarrassed asking questions in front of the class.

Using a math journal as a way to assess a student's progress (or lack thereof) allows you to see how students are interpreting and translating your instruction in their own words. You can check their comprehension and make sure nothing gets lost or misconstrued. If you collect the journals on a regular basis, it may help you to identify and help a student who is struggling.

Graphic Organizers

Graphic organizers are very helpful tools in adult education—especially when students can use them for their own purposes. There are graphic organizers for everything from note taking to problem solving. Students can even design their own. A good graphic organizer can provide a model for how students should think their way through a problem. Using a graphic organizer allows students to combine words and images to help them comprehend and remember concepts.

When you show a student how to use a graphic organizer, you give him a way to put his thought process on paper. Each student can use the same template, but no two will look alike. They are flexible and can be tailored to each student's needs. By giving students tools like these to support their learning, you also reinforce the idea that this time they will succeed.

One of the most-used graphic organizers is the Frayer Model. Many graphic organizers, like the Frayer Model, can be personalized to fit any situation. It's a great tool to use to help students represent their understanding of a new concept. This, and other graphic organizer templates, can be found in Appendix D. The Frayer Model can be used as a template for math journals.

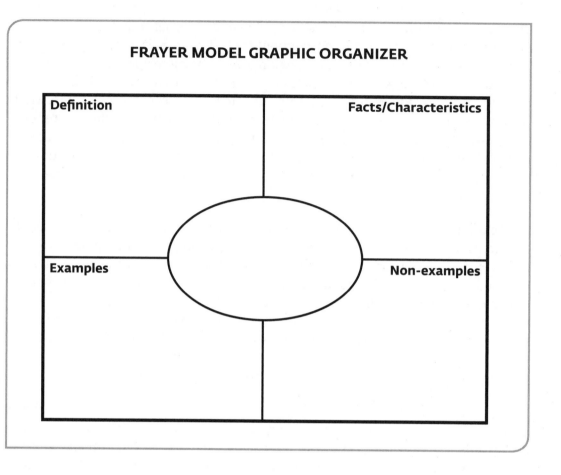

FRAYER MODEL GRAPHIC ORGANIZER

Definition	Facts/Characteristics
Examples	Non-examples

Manipulatives

Manipulatives are great tools for math instruction and exploration. Some teachers don't use manipulatives to fill the gaps in students' foundational math because they don't know how to use them. Others are very confident about using manipulatives, and they pass their excitement on to students. There are many types of manipulatives to choose from and many different applications. Here are some for adults learning math:

- Algebra tiles (my favorite manipulatives)
- Fraction packets (check out the video on making and using fraction packets)
- Cuisenaire® rods (pictured on next page)
- Dice and/or dominoes
- Styrofoam cups labeled with whole numbers
- Objects for counting
- Peel off 2D shapes

CUISENAIRE RODS

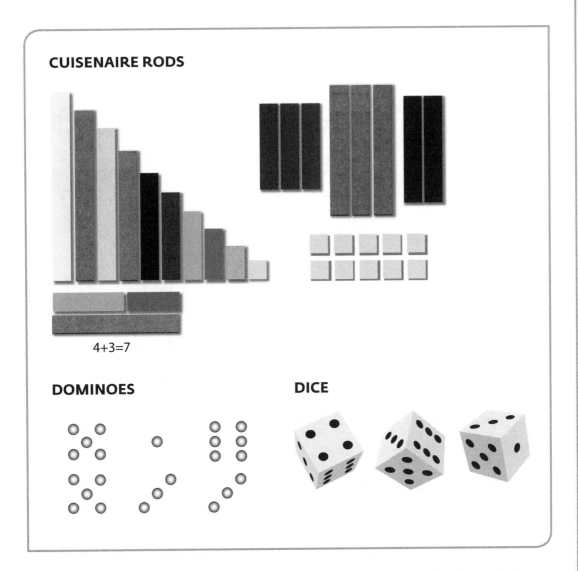

4+3=7

DOMINOES

DICE

The use of these manipulatives is demonstrated throughout the book and in the video activities that accompany this book.

If you cannot purchase manipulatives, you can make your own—or students can make them. See Appendix C for some templates. You can also search online for printable templates.

Mathematics Processes

The process—how the math is done and for what reason—is just as important as the content you teach. The term *process*, as used here, refers to the following:

- How math is taught (tools, strategies, and models, etc.)

- The way you encourage students to demonstrate their understanding of the math

- The multiple entry points from which students can begin to solve a problem

- Why you are assigning particular problems to students and what you expect them to learn from the problems

It is much simpler to correct errors students make while doing computations than to correct misunderstandings of fundamental mathematical concepts, but correcting fundamental errors is where the magic happens! You often hear parents complain that they can't help children with their homework because schools teach very different ways of computing solutions now than they did when parents were in school. It is true that adults learned to do math differently than the ways students are taught today. Instruction is now focused on building conceptual understanding before teaching shortcuts and computational algorithms.

As we think about the *how* of instruction, there are a great many things to consider. Both the National Council of Teachers of Mathematics (NCTM) and the writers of the CCRS begin their discussions of standards-based instruction with an emphasis on the skills and behaviors they want to see students develop as they do math. Process before content. Process is a big deal. Focusing on the *how* of our instruction is critical if we are going to see a change in our students' abilities to problem solve, reason, and make connections to the math they have done, are doing, and will do. This will be the area that will take the most time and concentration for instructors—bar none. Take a moment and acknowledge that instruction is a meld of many factors including the dynamics of a classroom. Think of the process part of your instruction as how you deliver your intended message.

Below, are the NCTM Process Standards and the Standards for Mathematical Practice, which are part of CCRS. These lists represent the processes that math standards developers value and the skills they encourage us to teach students.

NCTM Process Standards

Problem Solving: Solving problems is not only a goal of learning mathematics but also a major means of doing so. It is an integral part of mathematics, not an isolated piece of the mathematics program. Students require frequent opportunities to formulate, grapple with, and solve complex problems that involve a significant amount of effort. They are to be encouraged to reflect on their thinking during the problem-solving process so that they can apply and adapt the strategies they develop to other problems and in other contexts. By solving mathematical problems, students acquire ways of thinking, habits of persistence and curiosity, and confidence in unfamiliar situations that serve them well outside the mathematics classroom.

Reasoning and Proof: Mathematical reasoning and proof offer powerful ways of developing and expressing insights about a wide range of phenomena. People who reason and think analytically tend to note patterns, structure, or regularities in both real-world and mathematical situations. They ask if those patterns are accidental or if they occur for a reason. They make and investigate mathematical conjectures. They develop and evaluate mathematical arguments and proofs, which are formal ways of expressing particular kinds of reasoning and justification. By exploring phenomena, justifying results, and using mathematical conjectures in all content areas and—with different expectations of sophistication—at all grade levels, students should see and expect that mathematics makes sense.

Communication: Mathematical communication is a way of sharing ideas and clarifying understanding. Through communication, ideas become objects of reflection, refinement, discussion, and amendment. When students are challenged to communicate the results of their thinking to others orally or in writing, they learn to be clear, convincing, and precise in their use of mathematical language. Explanations should include mathematical arguments and rationales, not just procedural descriptions or summaries. Listening to others' explanations gives students opportunities to develop their own understandings. Conversations in which mathematical ideas are explored from multiple perspectives help the participants sharpen their thinking and make connections.

Connections: Mathematics is not a collection of separate strands or standards, even though it is often partitioned and presented in this manner. Rather, mathematics is an integrated field of study. When students connect mathematical ideas, their understanding is deeper and more lasting, and they come to view mathematics as a coherent whole. They see mathematical connections in the rich interplay among mathematical topics, in contexts that relate mathematics to other subjects, and in their own interests and experience. Through instruction that emphasizes the interrelatedness of mathematical ideas, students learn not only mathematics but also about the utility of mathematics.

Representations: Mathematical ideas can be represented in a variety of ways: pictures, concrete materials, tables, graphs, number and letter symbols, spreadsheet displays, and so on. The ways in which mathematical ideas are represented is fundamental to how people understand and use those ideas. Many of the representations we now take for granted are the result of a process of cultural refinement that took place over many years. When students gain access to mathematical representations and the ideas they express and when they can create representations to capture mathematical concepts or relationships, they acquire a set of tools that significantly expand their capacity to model and interpret physical, social, and mathematical phenomena. (NCTM)

Standards for Mathematical Practice:

1. **Make sense of problems and persevere in solving them:**
   Mathematically proficient adult learners start by explaining to themselves the meaning of a problem and looking for entry points to its solution. They analyze givens, constraints, relationships, and goals. They make conjectures about the form and meaning of the solution and plan a solution pathway rather than simply jumping into a solution attempt. They consider analogous problems, and try special cases and simpler forms of the original problem in order to gain insight into its solution. They monitor and evaluate their progress and change course if necessary. Higher level students might, depending on the context of the problem, transform algebraic expressions or change the viewing window on their graphing calculator to get the information they need. Mathematically proficient adult learners can explain correspondences between equations, verbal descriptions, tables, and graphs, or draw diagrams of important features and relationships, graph data, and search for regularity or trends. Lower level students might rely on using concrete objects or pictures to help conceptualize and solve a problem. Mathematically proficient students check their answers to problems using a different method, and they continually ask themselves, "Does this make sense?" They can understand the approaches of others to solving complex problems and identify correspondences between different approaches.

2. **Reason abstractly and quantitatively:**
   Mathematically proficient adult learners make sense of quantities and their relationships in problem situations. They bring two complementary abilities to bear on problems involving quantitative relationships: the ability to decontextualize—to abstract a given situation, represent it symbolically, and manipulate the representing symbols as if they have a life of their own, without necessarily attending to their referents—and the ability to contextualize (to pause as needed during the manipulation process in order to probe into the referents for the symbols involved). Quantitative reasoning entails habits of creating a coherent representation of the problem at hand;

considering the units involved; attending to the meaning of quantities, not just how to compute them; and knowing and flexibly using different properties of operations and objects.

3. **Construct viable arguments and critique the reasoning of others:**
Mathematically proficient adult learners understand and use stated assumptions, definitions, and previously established results in constructing arguments. They make conjectures and build a logical progression of statements to explore the truth of their conjectures. They are able to analyze situations by breaking them into cases, and can recognize and use counterexamples. They justify their conclusions, communicate them to others, and respond to the arguments of others. They reason inductively about data, making plausible arguments that take into account the context from which the data arose. Mathematically proficient adult learners are also able to compare the effectiveness of two plausible arguments, distinguish correct logic or reasoning from that which is flawed, and—if there is a flaw in an argument—explain what it is. Lower level adult learners can construct arguments using concrete referents such as objects, drawings, diagrams, and actions. Such arguments can make sense and be correct, even though they are not generalized or made formal until higher levels. Later, students learn to determine domains to which an argument applies. Students at all levels can listen or read the arguments of others, decide whether they make sense, and ask useful questions to clarify or improve the arguments.

4. **Model with mathematics:**
Mathematically proficient adult learners can apply the mathematics they know to solve problems arising in everyday life, society, and the workplace. In lower levels, this might be as simple as writing an addition equation to describe a situation. In intermediate levels, adult learners might apply proportional reasoning to plan a school event or analyze a problem in the community. By the upper levels, an adult learner might use geometry to solve a design problem or use a function to describe how one quantity of interest depends on another. Mathematically proficient adult learners who can apply what they know are comfortable making assumptions and approximations to simplify a complicated situation, realizing that these may need revision later. They are able to identify important quantities in a practical situation and map their relationships using such tools as diagrams, two-way tables, graphs, flowcharts, and formulas. They can analyze those relationships mathematically to draw conclusions. They routinely interpret their mathematical results in the context of the situation and reflect on whether the results make sense, possibly improving the model if it has not served its purpose.

5. **Use appropriate tools strategically:**
Mathematically proficient adult learners consider the available tools when solving a mathematical problem. These tools might include pencil and paper, concrete models, a ruler, a protractor, a calculator, a spreadsheet, a computer algebra system, a statistical package, or dynamic geometry software. Proficient adult students are sufficiently familiar with tools appropriate for their grade or course to make sound decisions about when each of these tools might be

helpful, recognizing both the insight to be gained and their limitations. For example, mathematically proficient advanced level adult learners analyze graphs of functions and solutions generated using a graphing calculator. They detect possible errors by strategically using estimation and other mathematical knowledge. When making mathematical models, they know that technology can enable them to visualize the results of varying assumptions, explore consequences, and compare predictions with data. Mathematically proficient students at various levels are able to identify relevant external mathematical resources, such as digital content located on a website, and use them to pose or solve problems. They are able to use technological tools to explore and deepen their understanding of concepts.

6. **Attend to precision:**
Mathematically proficient adult learners try to communicate precisely to others. They try to use clear definitions in discussion with others and in their own reasoning. They state the meaning of the symbols they choose, including using the equal sign consistently and appropriately. They are careful about specifying units of measure, and labeling axes to clarify the correspondence with quantities in a problem. They calculate accurately and efficiently, expressing numerical answers with a degree of precision appropriate for the problem context. In the lower levels, adult learners give carefully formulated explanations to each other. By the time they reach the advanced levels, they have learned to examine claims and make explicit use of definitions.

7. **Look for and make use of structure:**
Mathematically proficient adult learners look closely to discern a pattern or structure. Lower level students, for example, might notice that three and seven more is the same amount as seven and three more, or they may sort a collection of shapes according to how many sides the shapes have. Later, students will see 7×8 equals the well-remembered $7 \times 5 + 7 \times 3$, in preparation for learning about the distributive property. In the expression $x^2 + 9x + 14$, higher level students can see the 14 as 2×7 and the 9 as $2 + 7$. They recognize the significance of an existing line in a geometric figure and can use the strategy of drawing an auxiliary line for solving problems. They also can step back for an overview and shift perspective. They can see complicated things, such as some algebraic expressions, as single objects or as being composed of several objects. For example, they can see $5 - 3(x - y)^2$ as 5 minus a positive number times a square and use that to realize that its value cannot be more than 5 for any real numbers x and y.

8. **Look for and express regularity in repeated reasoning:**
Mathematically proficient adult learners notice if calculations are repeated and look both for general methods and for shortcuts. Intermediate level students might notice when dividing 25 by 11 that they are repeating the same calculations over and over again, and conclude they have a repeating decimal. By paying attention to the calculation of slope as they repeatedly check whether points are on the line through (1, 2) with slope 3, intermediate level students might abstract the equation $(y - 2)/(x - 1) = 3$. Noticing the regularity

in the way terms cancel when expanding $(x - 1)(x + 1)$, $(x - 1)(x^2 + x + 1)$, and $(x - 1)(x^3 + x^2 + x + 1)$ might lead them to the general formula for the sum of a geometric series. As they work to solve a problem, mathematically proficient adult students maintain oversight of the process, while attending to the details. They continually evaluate the reasonableness of their intermediate results.

If you're looking at the Standards for Mathematical Practice (SMP) for the first time, you will need to think about how to integrate them into your instruction. Here is one way that you can draw connections between the NCTM Process Standards and the SMPs. Using both the process and practice standards can help you develop the critical framework on which to build your instruction.

NCTM PROCESS STANDARDS & CCRS STANDARDS FOR MATHEMATICAL PRACTICE

	Problem Solving	Representation	Connection	Reasoning & Proof	Communication
Make sense of problems and persevere in solving them.	✔				
Reason abstractly and quantitatively.				✔	
Construct viable arguments and critique the reasoning of others.				✔	✔
Model with mathematics.	✔	✔			
Use appropriate tools strategically.	✔				
Attend to precision.	✔	✔			
Look for and make use of structure.			✔	✔	
Look for and express regularity in repeated reasoning.				✔	

Common Core State Standards for Mathematics, 2010

The driving force of the paradigm shift toward emphasizing process and content is the NCTM. Though the NCTM focuses primarily on K-12 students, a great deal of the research is applicable to adult education. And most math standards reference the NCTM standards. Looking at the adult education SMPs in relation to the NCTM Process Standards they fall under can help you recognize what process (*how*) strand or family to focus on for every lesson.

The crafters of the SMPs recommend that one lesson use no more than two to three of the SMPs to facilitate learning, and no more than three or four of the standards. Though it may be tempting to plan a lesson that checks off as many boxes as possible, try to resist that temptation. That is the old way of doing things.

Designing Rigorous Lessons

Crafting rigorous lessons that are challenging and engaging for students may seem a little different from what you are used to, but often you can tweak the lessons you currently use to meet all your instructional needs. Simply changing the questions you ask can deepen the experience for students, and challenge them to go a little further in their solutions.

Here are a few guidelines to follow when writing practice questions. According to Sullivan and Lilburn (2002), "There are three main features of good questions:

1. They require more than remembering a fact or reproducing a skill

2. Students can learn by answering the questions, and the teacher learns about each student from the attempt

3. There may be several acceptable answers."

Sullivan and Lilburn contend that questions with multiple answers stimulate students' "higher level thinking and problem solving."

It is tempting to follow the order of the content standards when planning lessons. It might seem easy to pick a couple of content standards, a couple of process standards, and/or SMPs, and then write a lesson plan. Remember that you need to ask deeper, more probative questions. You also need to focus on how to assess how much progress students are making. Then move on to the next standard.

Repeating this process mindlessly won't help prepare students for the rigorous demands of college and/or 21st century careers. Classroom planning needs to be purposeful. The key to crafting lessons that are rich and robust is relevance—teaching math that means something. At the higher levels, math becomes more abstract, and that is certainly valuable to students who are pursuing technology courses or careers. However, the instruction—the problems used to illustrate the concepts—and the exercises or projects assigned should still be relevant.

A great example of this is a lesson taught by my friend Brooke Istas, the moderator of the math and numeracy discussion group on LINCS. First she showed students a video clip of a family-run gelato business. Then she played it again. The second time, she asked her students to note all the instances where math was used. After being asked to pay attention to the math, they noticed math in every minute of the video. A very productive discussion ensued. Students realized that math is all around them all the time, even when they are not aware.

She then got into the heart of the lesson while covering several important objectives. She first posted five pieces of paper, each labeled with an ice cream flavor, on the walls in the classroom. She then asked students to choose their two favorite flavors of ice cream from a choice of five flavors (chocolate, vanilla, fruit, nut, and other). They marked their favorite flavors on the corresponding sheet on the wall. Then they tallied the total number of votes each flavor received, and made a bar graph on graph paper with each flavor represented by a different color.

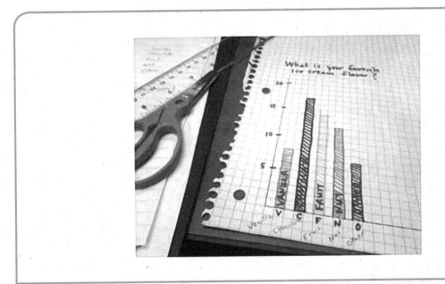

Next, they cut out the strips from the bar graph.

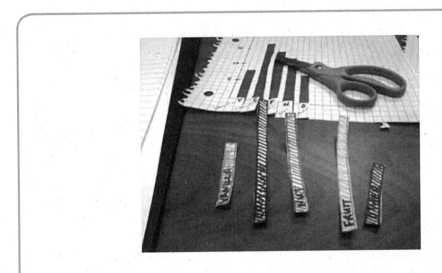

Then they laid the strips end-to-end and taped them together.

They connected the ends together, making a circle.

The next step was to trace the circle onto another piece of paper, marking the places where each colored bar began and ended.

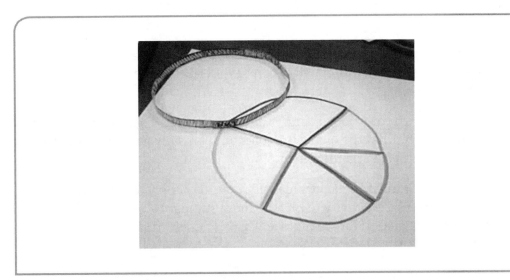

Finally, they labeled the pie charts with the percentages of total votes each flavor received.

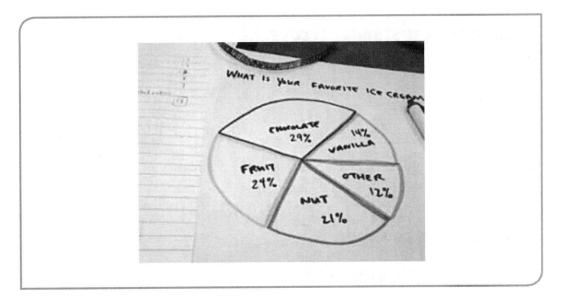

This is a multifaceted activity that you could use in many different ways. That is the hallmark of a really rich activity! As you reflect on the lesson and activity, which of the content standards do you think it addresses?

Possible CCRS content standards covered:

- **1.MD.4** Organize, represent, and interpret data with up to three categories; ask and answer questions about the total number of data points, how many in each category, and how many more or less are in one category than in another.

- **2.MD.10** Draw a picture graph and a bar graph (with single-unit scale) to represent a data set with up to four categories. Solve simple put-together, take-apart, and compare problems using information presented in a bar graph.

- **6.SP.2** Understand that a set of data collected to answer a statistical question has a distribution which can be described by its center, spread, and overall shape.

- **6.RP.3c** Find a percent of a quantity as a rate per 100 (e.g., 30% of a quantity means 30/100 times the quantity); solve problems involving finding the whole, given a part and the percent.

Are there any more standards you can identify that may apply to this activity? How could you extend the activity to include more standards? How would you differentiate the instruction for students doing this activity at a lower or higher level?

Consider prepping for and facilitating this activity: What types of skills or behaviors would you target? Think about which of the NCTM Process Standards apply to this activity.

Possible process standards covered:

- Problem Solving

- Connections

- Representations

Can you identify any others? Now think about the Standards for Mathematical Practice (SMPs). Which of those apply?

Possible SMPs covered:

- Make sense of problems, and persevere in solving them

- Reason abstractly and quantitatively

- Use appropriate tools strategically

Are there any others you would target or include?

Including tools like the content and process standards in your lesson planning will help you to build deep, coherent, and rigorous lessons. Thinking not just about the *what* of instruction, but also the *how* will help you become more confident and competent in designing your lessons.

Fewer but Deeper

To prepare your students for their journeys beyond the classroom, you do not have to teach every standard. Some of the content standards will be supportive or supplemental rather than core. To plan your instruction, first determine which standards your program will be expected to use. Is your adult education program following the College and Career Readiness Standards? It is important to know for which set of standards you will be held accountable. Also keep in mind other factors that will shape your program and its standards: What opportunities exist for your students as far as college and the workforce are concerned? What are the employment trends in your community? What do you know about the entrance exams for local colleges? These are all questions you should consider when designing lessons for your students.

The focus of the CCRS is to provide standards-based instruction most likely to lead to successful entry into college or the workplace, without the need for any remedial or developmental course work or training. The CCRS do not include the low ABE domain of counting and cardinality, or the ASE domains of the complex number system and vector and matrix quantities. Why does this matter?

Most students who matriculate from a traditional high school education—having taken four years of math—will have some experience doing math that includes complex numbers and vectors (usually in Algebra II, Physics, or both). But entrance exams for colleges do not require that level of mathematics. Also adult education students generally spend much less time in class than do high school graduates. The CCRS were developed to prepare students to enter college without having to take developmental classes. Since students will not encounter complex numbers and vectors unless they take additional math classes in college, the corresponding standards were deemed less important. The same is true for most trade careers and workplace requirements.

Regarding the lowest domain—counting and cardinality—I do not know the rationale for excluding the domain. I can only guess that the standards developers assumed that students on a path toward college or careers would already have basic numeracy skills.

Skills Needed for Successful Transition to College and Careers

How do these standards relate to what adult learners need for transition to college or careers? According to a recent study conducted by Hart Research Associates for the Association of American Colleges and Universities, colleges and employers see the greatest indicators for long-term success as developing and leveraging a broad range of skills. Many students want to focus their time with us on temporarily "learning" the math they need to pass the HSE test they take, without appreciating the need to learn how to think critically or problem solve. This is why the processes we lead our students through as we deliver our instruction are so crucial (Hart Research Associates, 2015).

So, what skills do students need to successfully transition to college and/or careers? Take a look again at the NCTM Process Standards and the CCRS Standards for Mathematical Practice on pages 71 through 75. These standards and practices are skills that students can use in college, at the workplace, and in everyday life.

Math Operations

Having a command of numbers and their relationships is not the whole of math. It's also important to understand the vocabulary of math. It might be helpful to keep a running list of math vocabulary terms or a word wall in the classroom where students can see it. Another way to deal with the vocabulary terms is to incorporate them in students' math journals. Have them write new terms in their journals and include definitions and examples. Then they can refer to their math glossaries whenever they need to.

Numbers

A relationship with math begins with numbers. Everyone interacts with math every day, sometimes without even realizing it.

Numbers are the foundation for many things people do:

- If you want to cook dinner, you need to know how much food to make to serve a certain number of people. And that may require using measurements to follow a recipe.

- When you go to a movie, you need to subtract the cost of tickets from the money in your wallet to see what size popcorn you can afford.

- You read an international news article that uses metric numbers. How can you relate to the temperatures being forecasted? How do you know how far a kilometer is? How does the value of another country's currency compare to yours?

- Your child has a fever, so you take his temperature using a thermometer. What does the number mean? How high is too high?

- How many times can you hit the 9-minute snooze button on your alarm clock before you really need to get out of bed in the morning?

The foundation for math begins with knowing what numbers are, what they represent, and how they can be used. The CCSS calls this introductory content "Counting and Cardinality." The CCRS do not include this level, because they assume that adult students already have a conceptual understanding of the most basic math content. So, assuming that adult learners are familiar with numbers, start with place value. This is an important concept to master, since subsequent skills such as math operations and working with decimals depend on mastery of place value.

First, make sure all students are on the same page with a quick review of how ones, tens, hundreds, and other place values are related.

Write a number on the board and ask "What is this?"

Responses will vary:
- "A really big number!"
- "Nine hundred and sixty two"
- "Almost 1,000"

Let students continue to think about this for a few minutes, and encourage them to think deeper by adding, "Yes, and what else?"

Some students will get there on their own, but others need to be led. You are waiting to hear someone say, "Nine hundreds, six tens, and two ones."

Understanding place value is crucial to success in math. Without a sense of place value, numbers and operations remain a mysterious, confusing part of the universe that only some people understand.

Place Value and Beginning Operations

The best way to teach or review place value is with manipulatives, such as base ten blocks, a place value mat and chips, 10-frames, or algebra tiles.

Base Ten Blocks

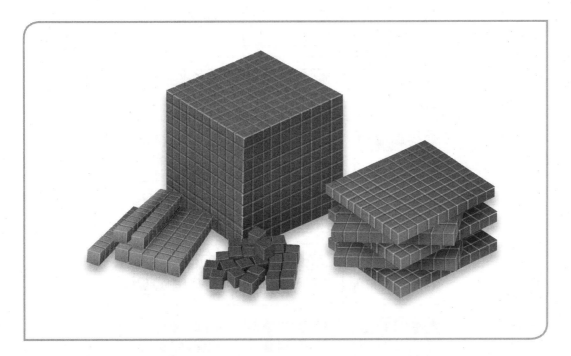

Place Value Mat and Chips

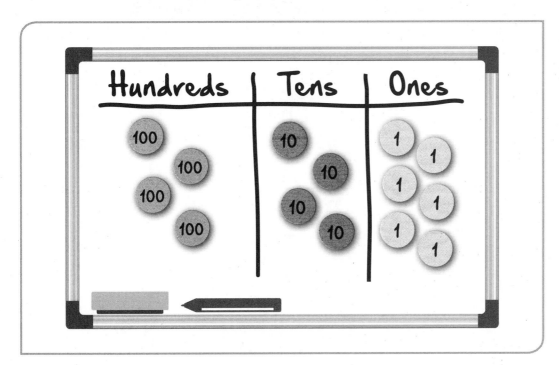

Ten Frames

Ten frames can either be concrete or representational. They allow students to practice combining numbers to make 10—a crucial skill in building number sense.

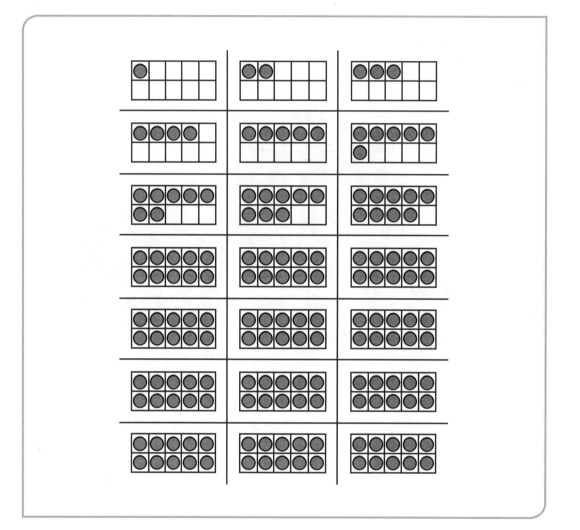

Algebra Tiles

Algebra tiles can be used to model many mathematical concepts, including integers.

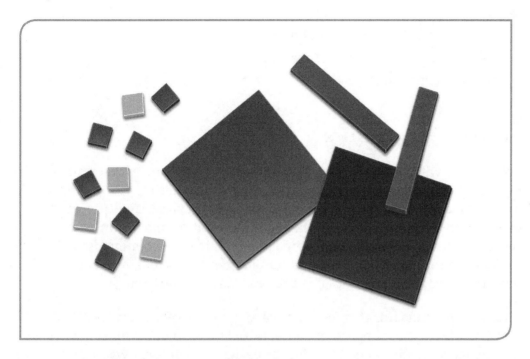

Base ten blocks, place value mats, ten frames, and algebra tiles can help many students start building their mathematical competencies.

If you cannot purchase manipulatives, you can make your own—or students can make them. See Appendix C for some templates. You can also search online for printable templates.

Hands-on Practice

There are many examples that will help you to get the most out of using manipulatives, as well as transition to using the tools to demonstrate performing operations or calculations. Keep in mind that you need to spend enough time for students to master one concept before moving on to another. Take the time your students need to not only practice the concepts, but to make students comfortable enough that they no longer need to rely on the manipulatives to perform the tasks. If your students can move from the concrete manipulatives to drawing models, they are ready to move on to more abstract concepts. This will set them up for success in applying their previous understandings to more sophisticated math.

Operations

Before starting with operations, you have to be certain students have a firm grasp of numbers and what they represent. Build on that foundation to begin looking at the rest of the domains in this standard:

- Operations and Algebraic Thinking
- Numbers and Operations in Base 10
- Measurement and Data
- Geometry

Look at this standard in CCRS Level A:

1.NBT.4 Add within 100, including adding a two-digit number and a one-digit number, and adding a two-digit number and a multiple of 10, using concrete models or drawings and strategies based on place value, properties of operations, and/or the relationship between addition and subtraction; relate the strategy to a written method and explain the reasoning used. Understand that in adding two-digit numbers, one adds tens and tens, ones and ones; and sometimes it is necessary to compose a ten.

This is one of the standards that focus on the Major Work of the Level (MWOTL). The MWOTL are the most important concepts and skills students must master before they move on. Check the CCRS for the designation of the MWOTL and supporting standards. For more information on MWOTL for CCRS math, go to: https://lincs.ed.gov/programs/ccr/math

 Video Activity: Operations with the Place Value Mat and Chips

Break instruction for this content into a couple of different categories: models and strategies. Models, such as graphs, charts, diagrams, and equations, help you make sense of mathematical situations. Strategies are the problem solving steps used or the ways of approaching a problem to solve it. Before you begin to teach the content in this standard, understand the following:

- You are working with numbers up to 100.
- One- and two-digit numbers will be added together.
- You will be using concrete or representational models.
- You can incorporate place value and properties of operations.
- You can include the relationship between addition and subtraction.

Addition

MODEL: 28 + 7 =
STRATEGY: ADDING ON AN OPEN NUMBER LINE

Students use a number line and decompose addends to represent the operation of addition. In this example, the number 7 is decomposed to 2 + 5 in order to easily represent it on the number line.

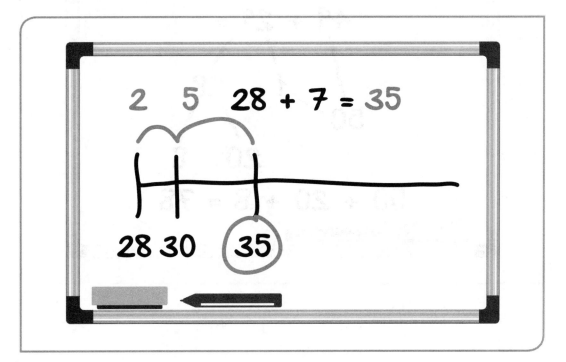

MODEL: 49 + 29 =
STRATEGY: MAKING A *NICE* NUMBER

Students decompose and regroup numbers until they have *nice* (workable) numbers to use.

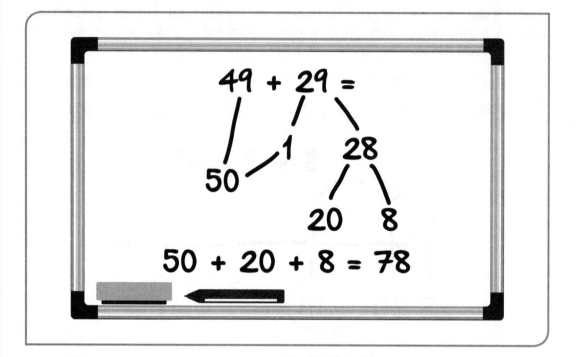

MODEL: 38 + 17 =
STRATEGY: HORIZONTAL ADDITION, PART 1

Students decompose and regroup numbers another way to perform addition.

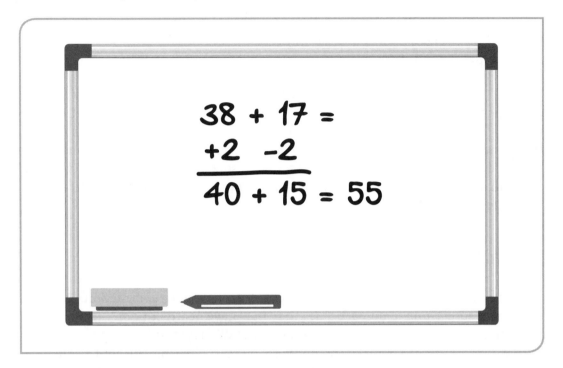

MODEL: 38 + 17 =
STRATEGY: HORIZONTAL ADDITION, PART 2

Students add the tens and the ones separately, and then combine them, regrouping if necessary.

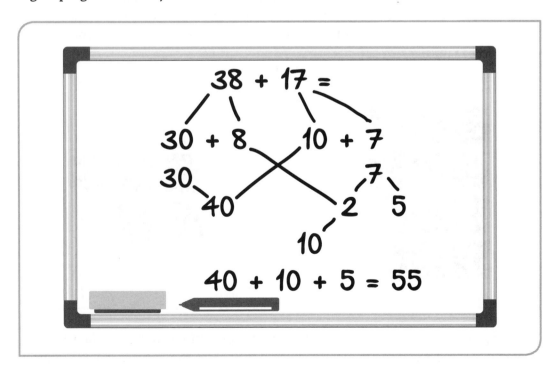

Use these strategies one-to-one or in large or small group instruction. Emphasize "think-alouds" while you present. As an example, talk about your thought process as you demonstrate what you are doing. This can include asking—and answering, if necessary—questions like these:

- What would make these numbers easier to work with?
- How can I regroup these numbers so I can add them together?
- Which number should I decompose, and which should I keep whole?

Students need some basic number sense before they will be able to employ these strategies on their own, so it is crucial to work on building those skills first. In the early levels of math, make it a practice to ask students, "What would make these numbers easier to use?" With enough practice, students will remember to try these strategies when they work on problem solving. The concept of breaking a problem into manageable chunks works even when the math is more abstract and sophisticated.

Subtraction

Some students who do well with addition may struggle with subtraction. This could have something to do with rules they may have heard but don't really understand, especially with integers. Rules like, "When you subtract, the answer will always be less than the number you started with." Without a solid understanding of integers, rules like this can be more confusing than helpful.

MODEL: 12 – 7 =
STRATEGY: SUBTRACTION WITH ALGEBRA TILES

Algebra tiles can be used to model subtraction. Start with the problem of subtracting 7 from 12. Lay out 12 small cubes from the algebra tiles, and then subtract 7 from the group. Count what is left, and that is the answer to the question of 12 minus 7, or 12 – 7 = ___.

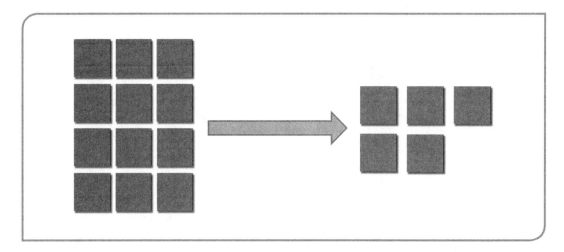

Modeling subtraction with algebra tiles might seem simple, but the goal is for students to be very comfortable with these skills before they transition to higher

TEACHING ADULTS: A MATH RESOURCE BOOK

math. It also helps to build confidence when students can find their own way to solve simple problems using the tiles.

The same strategies we used for addition can be used for subtraction: open number line, making a *nice* number, and horizontal subtraction.

 Video Activity: Operations on an Open Number Line

MODEL: 37 – 9 =
STRATEGY: SUBTRACTION ON AN OPEN NUMBER LINE

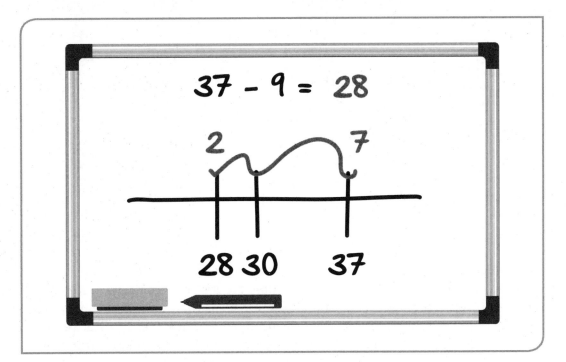

 Video Activity: Using Ten Frames

 Video Activity: Using Base Ten Blocks

Multiplication

There are many ways to think about multiplication, from how it is represented to how it is performed. There is the classic algorithm that many people were taught, and there is also lattice multiplication. But, one of the best ways to fill gaps in your students' understanding of what's going on in the multiplication process is to use array multiplication.

Array multiplication is simply arranging equal groups of items in rows and columns. Here are two different pictures of a grouping of 6 oranges:

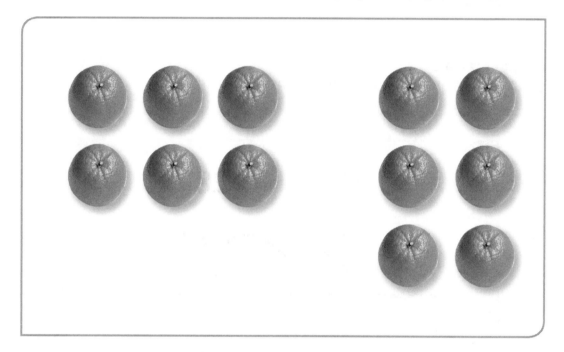

The picture on the left can be interpreted as "two groups of three oranges." Likewise, the picture on the right shows "three groups of two oranges." In each picture, however, there are only six oranges. The ideas that need to be conveyed here are important, but do not necessarily need to be named (i.e., associative property). Here are the concepts students need to understand and own from this point on:

- The operation of multiplication shortens the process of repeated addition.
- The order of the multiplicands can change, but the product remains the same.

MODEL: 216 × 32 =
STRATEGY: LATTICE MULTIPLICATION

Lattice multiplication is a strategy to multiply larger numbers using a grid. You need one row or column for each digit in the two multiplicands. In this case, three columns represent a 3-digit number and 2 rows represent a 2-digit number.

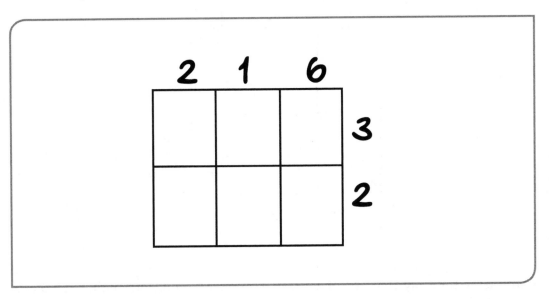

After you make the grid, draw diagonal, dotted lines to divide each rectangle in half.

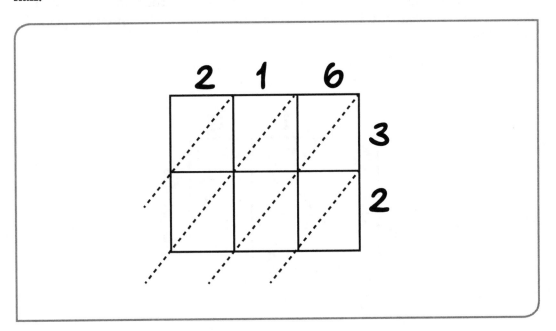

Next, begin multiplying. Start with the ones position. Multiply 2 × 6. The product has two digits, so put the tens digit above the dotted line and the ones digit below it.

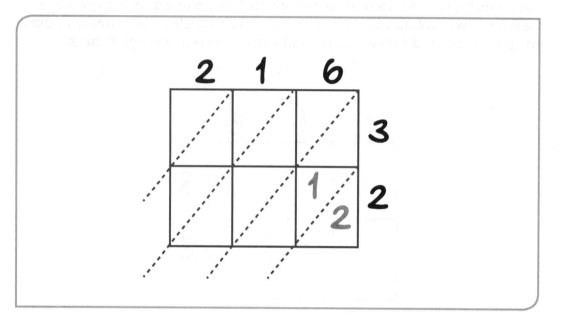

Continue to multiply each pair of digits. If a product has only one digit, put a 0 in the tens place, above the dotted line. Since you are multiplying one digit by one digit at a time, you will never have more than a two-digit product. When you have filled in the grid, start in the bottom-right corner, and add the numbers in each diagonal line. Notice that 8 + 1 + 2 = 11, so you need to carry the 1 into the next line: 1 + 1 + 3 + 0 + 4 = 9.

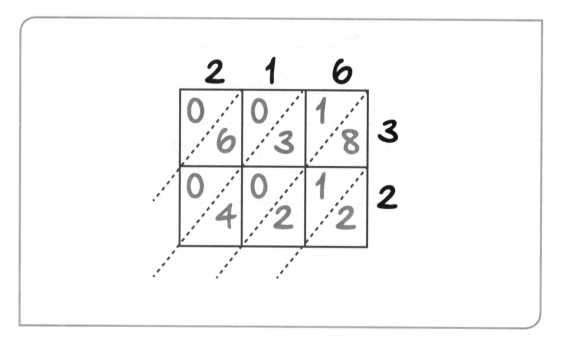

Answer: 6,912

Now compare lattice multiplication to the standard algorithm most people are used to, and you can see how this tweak to a multiplication array works to multiply 216 by 32. In the standard algorithm, we would first multiply each of the digits in 216 by 2, and record the products below the multiplicands. Next, we would offset the second line of products, and multiply each of the digits in 216 by 3. Finally, we would add the two lines of products together.

This is just a procedure, and it is likely that if you were to ask your students what is occurring, they would struggle to answer. For example:

- Why is the second line of the products indented?
- Why do we start multiplying from the right when we read from the left?
- Why do we also add from the right when we have all the products computed?

$$
\begin{array}{r}
216 \\
\times\ 32 \\
\hline
432 \\
+\ 648 \\
\hline
6912
\end{array}
$$

MODEL: 216 × 32
STRATEGY: PARTIAL-PRODUCT MULTIPLICATION

Here is yet another representation of what is happening in the multiplication problem. This is the partial product method. This method helps students see what is really happening in multiplication with respect to place value.

Using the 32, start with the ones column and multiply the 2 by all the digits in the three-digit multiplicand; this equals the 432 from the top representation because you are really multiplying

$$
\begin{aligned}
2 \times 6 &= 12 \\
2 \times 10 &= 20 \\
2 \times 200 &= 400
\end{aligned}
$$

Then move to the tens column and multiply the 30 by all the digits in the three-digit multiplicand:

$$
\begin{aligned}
30 \times 6 &= 180 \\
30 \times 10 &= 300 \\
30 \times 200 &= 6000
\end{aligned}
$$

$$
\begin{array}{r}
216 \\
\times\ 32 \\
\hline
12 \\
20 \\
400 \\
180 \\
300 \\
+\ 6000 \\
\hline
6912
\end{array}
$$

The partial products model breaks down the math more clearly to show how it leads to the product.

When you see these two models—the standard algorithm and the partial products method—side by side, it's obvious why some students don't understand the standard algorithm. It's a truncated system that actually hides or skips a few steps.

If your students appear to be struggling with any of these methods, go back to the beginning, and use the place value mat and chips to represent a multiplication problem. Use a three-digit number, like 216, and choose a single digit for the other multiplicand, like 3.

MODEL: 216 × 3 =
STRATEGY: PLACE VALUE MAT MULTIPLICATION

First divide the Place Value Mat into 3 sections.

Using chips, represent the number 216 three times on the mat—in three rows.

Next, examine the mat to see if any columns of chips need to be regrouped. For example, look at the ones column: There are 18 one chips in the column. These can be regrouped into 1 ten chip and 8 one chips.

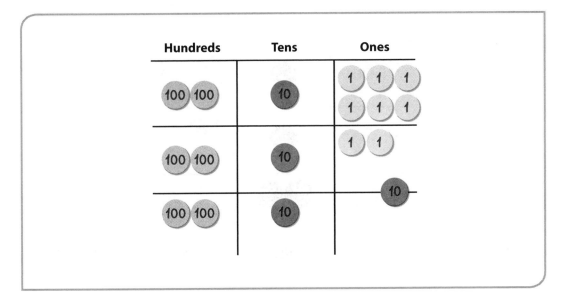

Move the new ten chip to the tens column.

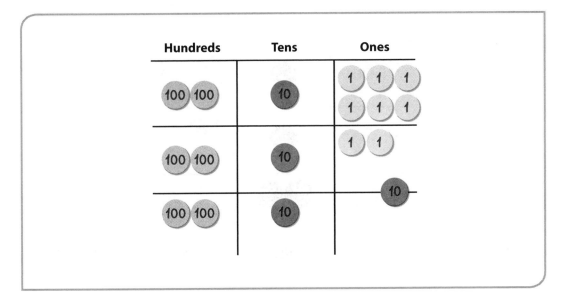

And finally, recompose or add the chips in each column to find the product of 216 × 3:

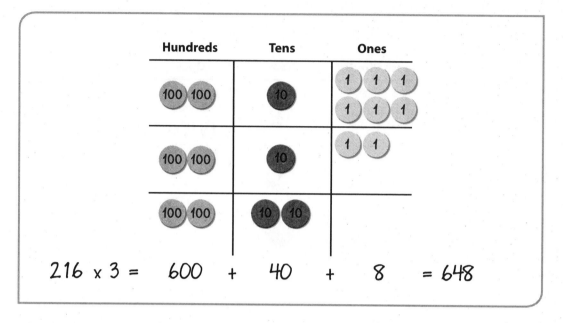

$$216 \times 3 = 600 + 40 + 8 = 648$$

The product of 216 and 3 is 648. And you can say that 216 and 3 are both factors of 648.

 Video Activity: Multiplication Strategies

Division

When you introduce a new topic, such as division, begin by familiarizing students with the vocabulary terms, e.g., *dividend, divisor, quotient, remainder*. Make sure students understand what the terms mean so that they can use them when they talk about solving division problems.

Division can look similar to multiplication, especially when using manipulatives to introduce the concept to students. Division starts with grouping.

Ask students: "How would you divide 36 by 9?" Explain it this way: "You have 36 objects, and you want to split them into 9 groups."

Students can use any kind of objects or counters you have available for this. If you start with the concrete manipulatives, your students should make 9 piles, starting with one or two counters in each pile. They may not know how many counters to put in each pile at this point, and that is fine. You may want to guide students by talking about your thought process aloud while your students are working. Try probing to see if they can estimate about how many counters should be in each pile.

Once students have made 9 piles with 4 counters in each pile, remind them of how they used the place value mat and chips for multiplication. (They divided the mat into three horizontal sections—because one multiplicand was 3, and they put 216 chips in each row—to represent the other multiplicand).

Ask students how and why partitioning their chips or counters in division is similar to what they did when they multiplied 216 × 3. You're looking for them to use vocabulary words, such as *factors*, *dividend*, *divisor*, and *quotient*. This is a great opportunity to have students write in their math journals and define the vocabulary in their own terms. You can also ask them to illustrate some division problems in their journals.

Once your students have created their piles and recorded their findings, ask them to identify what other factor, when multiplied by 9, equals 36. You want students to realize that the number of counters or objects in each pile is the other factor. This is a good way to check their number sense and see if they are understanding the concepts you're teaching.

 ## Video Activity: Division Strategies

Integers

If any of your students struggle with subtracting negative numbers, here is a way you can help them to grasp the concept. It begins with zero pairs. On page 92, we used algebra tiles to model subtracting 7 from 12. Look at that process again. Now we will subtract –7 from 12.

MODEL: 12 – (–7) =
STRATEGY: SUBTRACTING INTEGERS WITH ALGEBRA TILES

The problem looks like this: 12 – (–7) = ____

Start with 12 green cubes, but this time arrange them a little differently.

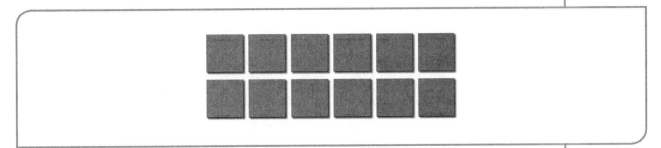

If you try to subtract –7 from these 12 cubes, you can't. You don't have any negative cubes. So, first add 7 negative cubes (literally). This is where zero pairs come in. You can't just insert random cubes into the problem, so for each negative cube that is added, you must add a positive one as well. Each positive and negative pair of cubes is called a zero pair, because the pair adds nothing to the total value of the cubes.

Now you still have 12 green (positive) cubes, but you also have 7 more green (positive) cubes and 7 gray (negative) cubes.

Subtract the 7 negative cubes.

To finish the problem, count the remaining green cubes. 12 – (–7) = 19. I chose to show this simple subtraction problem first, because this is one of the best strategies I have seen for helping students who struggle with integers. Students can see exactly what is happening when they add the zero pairs and subtract the negative number. Later, I will demonstrate how zero pairs also play a role when using algebra tiles to factor polynomials.

MODEL: 12 – (–7) =
STRATEGY: ADDING AND SUBTRACTING INTEGERS WITH A NUMBER LINE

You can also use a number line with integer problems.

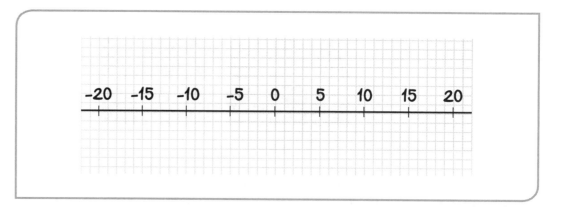

Just like word problems, these problems have a start (beginning number), a change (operation), and a result (answer).

Normally, when using a number line to demonstrate computations, the operation sign tells you which way to move (left or right) to find the answer. When you are adding positive integers, you move to the right from the starting point.

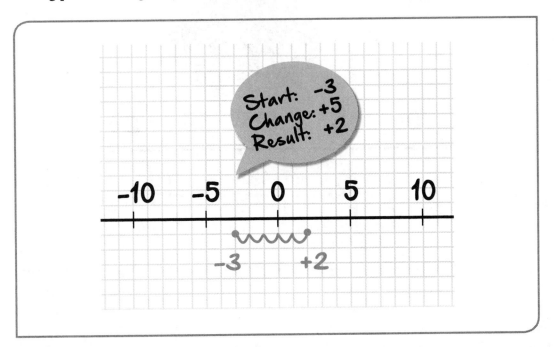

When you are subtracting positive integers, you move to the left from the starting point.

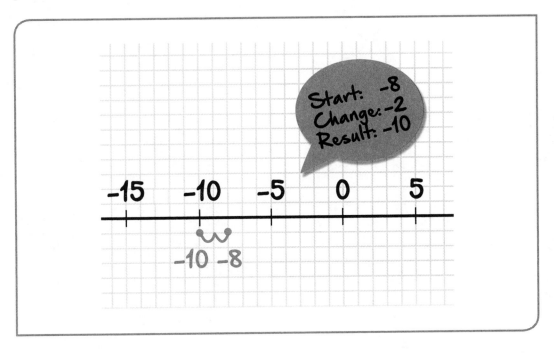

How can you use the number line to help students conceptualize the subtraction problem 12 – (–7) = ____?

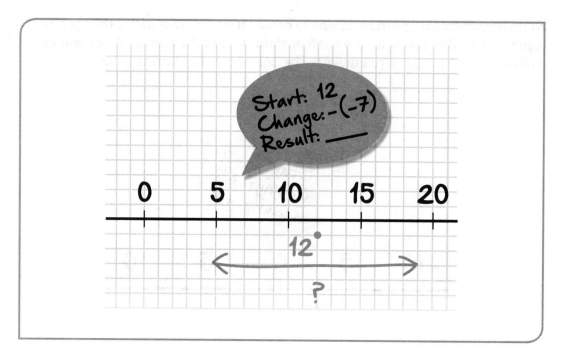

In the case of subtracting a negative number, the number makes us look at the operation differently. Think about subtraction as the distance between two numbers—or two points on the number line—and the measure of a distance is positive. If you plot the two points, 12 and –7, on a number line, the distance between them is 19. Both the algebra tiles and the number line are good visual demonstration tools to show students why subtracting a negative number results in a larger number.

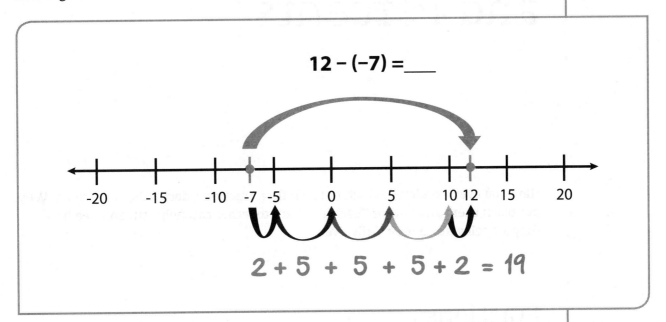

Fractions, Decimals, and Percents

Have you ever wondered which to teach first: fractions, decimals, or percents? Why not teach them all at once? Teaching them together can help students see how they are relevant to their daily lives.

Fractions

Begin by introducing fraction vocabulary terms, such as, numerator, denominator, and unit fraction. This will help students to understand and talk about fractions. Make sure they know the vocabulary before you start working on operations.

I recommend using fraction packets to teach students equivalent fractions. Each packet contains a series of equivalent fraction strips. See Appendix B for instructions on making and using fraction packets. You may also watch the video activity on making and using fraction packets.

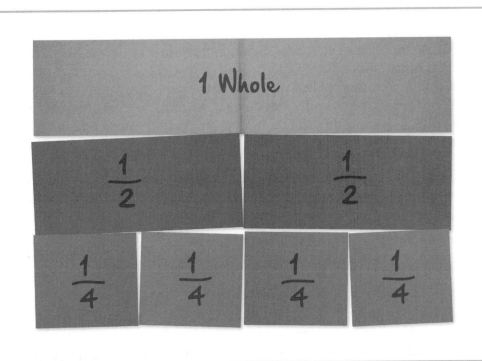

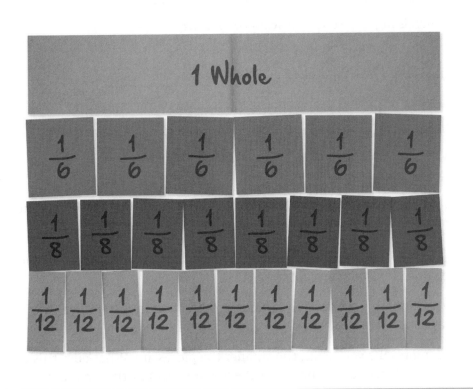

Students who struggle with fractions, decimals, and percents, will benefit from creating these tools. They should talk about them in class with other students. Don't make this a homework project or create the strips yourself. Students will benefit from the experience of creating the fraction packets in class.

To make the packets, use construction paper that is cut into 3" x 18" strips. These pieces seem easy for students to work with. I don't recommend laminating the strips, because the plastic edges prevent students from placing the fractions close together. Encourage students to be as precise as possible when folding and cutting the strips. It's important to make equal amounts exactly the same size. I do not make fifths or tenths, because it is not easy for students to approximate those sizes without a ruler. If you want to use them, you can find pre-made fraction strips at a teacher supply store or online.

When students are constructing these packets, ask some very specific questions to guide them. Ask questions to engage students and get them thinking about the fractions while they cut and mark the strips. For example:

- Once you have a $\frac{1}{2}$ fraction, how can you make a $\frac{1}{4}$ fraction?

- What do you notice as we are cutting the strips?

- Is $\frac{1}{3}$ of everything exactly the same size?

- What happens to the size of the fraction as the denominator gets larger?

When the fraction packets are completed (you can judge for yourself how many fraction strips you want to include in a packet), ask questions that encourage students to manipulate and compare the strips. For example:

- Which is bigger, $\frac{3}{4}$ or $\frac{5}{6}$? How do you know?

- Who has another "Which is bigger" question?

- What other fractions are the same length as $\frac{1}{2}$?

- Can you find other groups of fractions that are the same length?

- What do you know about fractions with the same denominator?

- What are you doing when you lay the fraction strips side by side? (Adding them)

- How can you compare the sizes of combined fractions? (Lay them out, one fraction under the other, compare which is longer/shorter/equal.)

Finish the activity by encouraging students to share their observations. Prompt them with questions or comments like these:

- What does this activity tell us about adding fractions together? Subtracting fractions?

- The more cuts you make, the smaller the pieces are. What does this tell you about fractions? (The larger the denominator is, the smaller the unit fraction is.)

- Two pieces can only be compared if they were cut from the same size whole. A fourth of a unit could be larger than a half unit depending on the original whole.

After students have used the fraction strips for a little while, and they have had a chance to make connections about unit fractions and equivalent fractions, you can play a game. This game is all about computations with fractions that have different denominators. For the game you will need one die per person and six small circle stickers for each die. Instructions for what to put on the stickers and how to play the game are in Appendix B.

Students can now use their fraction strips to complete worksheets with fraction questions. Here are some sample questions.

Use your fraction strips to answer the following questions:

1. How many eighths equal one whole?

2. What is $\frac{1}{2} + \frac{1}{4}$?

3. What is $1\frac{1}{4} - \frac{3}{8}$?

4. What is $2\frac{1}{3} \times \frac{1}{2}$?

5. What is $\frac{1}{4} \times 3$?

6. What is $\frac{3}{4} + \frac{3}{8}$?

*Credit for the Fraction Packet activity and instructions go to Donna Curry and Pam Meador.

To teach fractions, decimals, and percents at the same time, I like to put representations of currency on the back of some of the fraction strips. You can work simultaneously in fractions and decimals with $\frac{1}{2}$ and $\frac{1}{4}$. Begin with what I call *equivalence of value*. Equivalence is simple to explain: What is the decimal equivalent of 1 whole if we are using currency? ($1.00). Continue to ask questions, requiring students to connect fractions and money. What is the equivalence of value of $\frac{1}{2}$, if the whole equals $1.00? ($0.50). What is the equivalence of value of $\frac{1}{4}$ of one dollar? ($0.25). Knowing the equivalence of value for some common dollar amounts will also help students with word problems later on, when estimating whether an answer is reasonable or not.

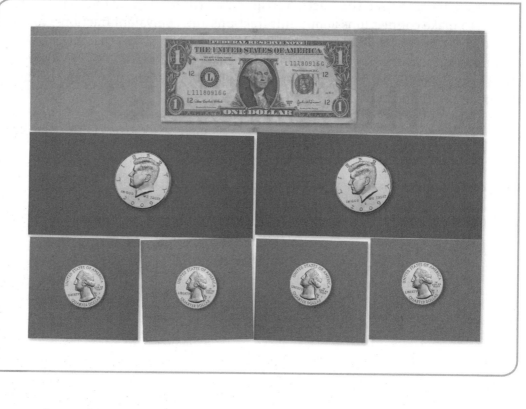

Working with Fractions

Now that students are comfortable working with fraction packets, they can move on to operations with fractions. Let's take a close look at CCRS 4.NF.2:

Compare two fractions with different numerators and different denominators, e.g., by creating common denominators or numerators, or by comparing to a benchmark fraction such as ½. Recognize that comparisons are valid only when the two fractions refer to the same whole. Record the results of comparisons with symbols > , = , or < , and justify the conclusions, e.g., by using a visual fraction model.

Problems or assignments that address this standard must contain fractions with unlike numerators and denominators—not students' favorite fractions. However, it may be easier for students to understand the concepts if you start by using fractions with like denominators. When students are comfortable working with like fractions, then you can move on to fractions with different denominators. You can start with fraction bars.

Activity 1

Adding Fractions with Like Denominators

PROBLEM: $\frac{3}{5} + \frac{1}{5} =$

Step 1: Create two fraction bars the same size. Divide each into five sections to represent the denominator of each fraction.

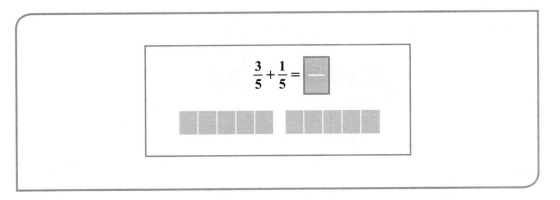

Step 2: Color in sections of each bar to represent the numerator of the fraction.

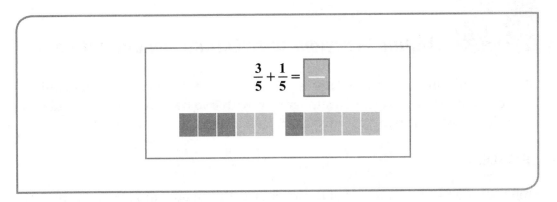

Step 3: Transfer the fewest number of boxes from one fraction bar to the other. In this example, move the representation of $\frac{1}{5}$ since there is only one unit to move.

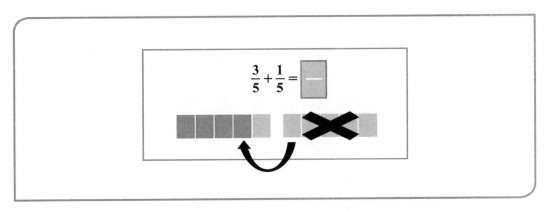

Step 4: When all the units from one fraction are moved, put an X over the fraction bar so students see that we are done with that fraction. This will help them to understand why we do not add the denominators. There is nothing left of that fraction to use.

Step 5: Now count all the colored units to get the numerator of the answer: 4. The total number of units in the fraction bar is the denominator: 5.

$$\frac{3}{5} + \frac{1}{5} = \boxed{\frac{4}{5}}$$

Answer: $\frac{3}{5} + \frac{1}{5} = \frac{4}{5}$

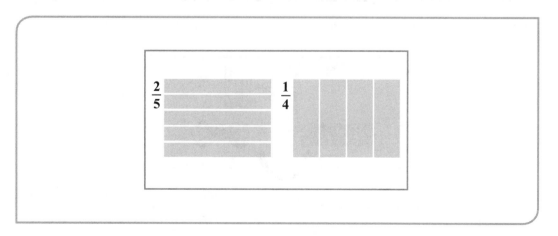

Activity 2

Adding Fractions with Unlike Denominators

When students understand the process of adding like fractions, move on to unlike fractions. This process starts in a similar way, with two fraction bars. Ask students, "How do we combine two fractions that do not have the same terms?

PROBLEM: $\frac{2}{5} + \frac{1}{4} =$

Step 1: Create two fraction bars of the same size. Divide one into five horizontal rows to represent the fraction with the denominator 5. Divide the other into four vertical columns to represent the other fraction with denominator 4.

Step 2: Next, color in the fractions to represent the numerators of both fractions.

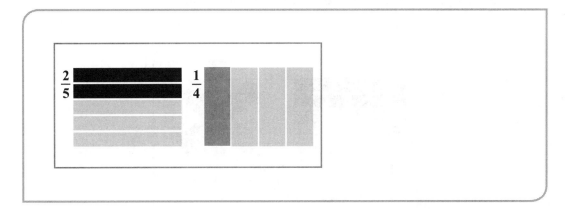

Step 3a: Copy the vertical column lines from the second fraction bar onto the first fraction bar.

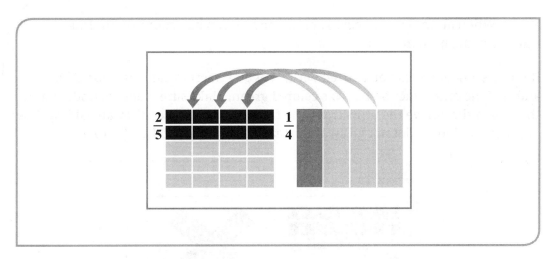

Step 3b: Copy the horizontal row lines from the first fraction bar onto the second fraction bar.

Looking at both of the fraction bars in this frame, you can see that we have the same number and size of units in each of them—a "common" denominator" has been created. You know this by counting the number of units in each fraction bar.

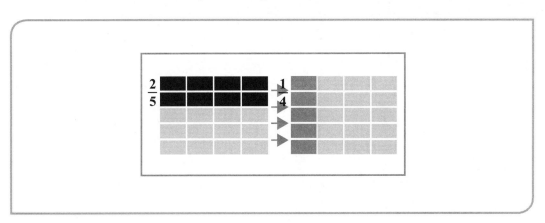

Step 4: Finally, move all the colored boxes onto one fraction bar. Put an X over the now empty fraction bar to signify that we are done with it.

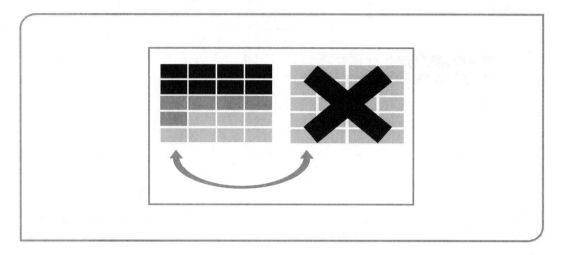

Step 5: When all the colored boxes are in one fraction bar, count them. This number is the numerator: 13.

Remind students what you started with: ⅖ of one fraction bar was colored black, and ¼ of the other fraction bar was colored green. Make sure students understand that when they move the units from one fraction to the other, they are adding. The original fraction (⅖) doesn't go away, instead they are just adding (¼) to it.

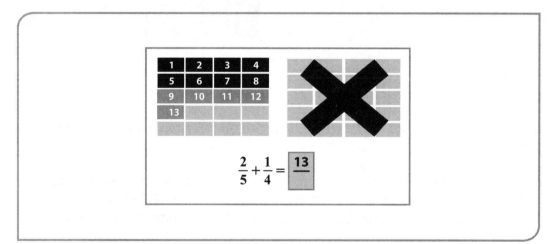

Step 6: Count the total number of boxes in the fraction bar. This is the denominator: 20. Now write the equation, and reduce the answer if possible.

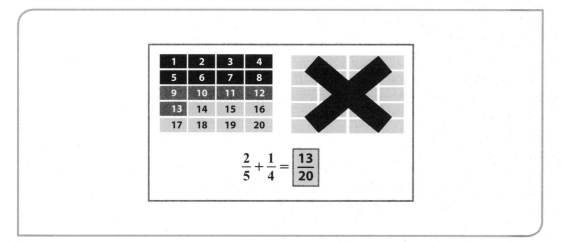

Answer: $\frac{2}{5} + \frac{1}{4} = \frac{13}{20}$

Suggestions: If you have a whiteboard or a chalkboard, use prepared templates of laminated fraction bars and units along with pre-cut pieces of painter's tape to demonstrate this on a large scale. Have students help to create the fraction bars. This gives kinesthetic learners a chance to interact with the model and get a good sense of what is really happening.

Activity 2 illustrates the *why before how* concept. When you do this activity, do not mention the words *common denominator*. Lean on part-whole thinking when you demonstrate these steps. You want students to imagine that they are creating a situation where both fraction bars have the same size and number of boxes. Work the word *common* into your classes by talking about "having an agreement" between the two fraction bars. If two people have something in common, like the kind of books or music they like, that means they agree about something. Expand that description of the word *common* by having students discuss it in groups or as a class.

Subtracting Fractions with Unlike Denominators

When you begin subtracting fractions, start your instruction the same way you started when adding fractions with different denominators. You may decide to start slowly, by using fractions with common denominators.

PROBLEM: $\frac{2}{3} - \frac{2}{5} =$

Step 1: Create two fraction bars of the same size. Divide one into three horizontal rows to represent the denominator 3. Divide the other into five vertical columns to represent the fraction with denominator 5.

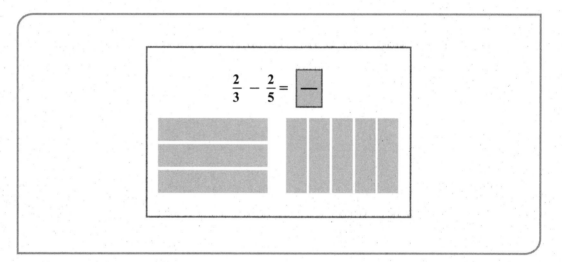

Step 2: Next, color in the fractions to represent the numerators of both fractions.

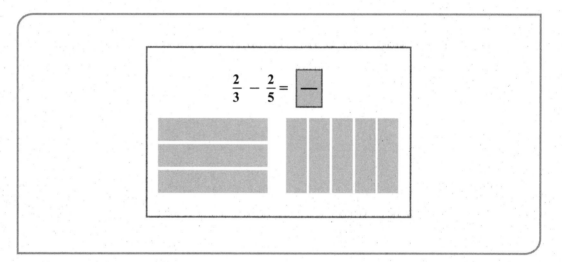

Step 3: Copy the vertical column lines from the second fraction bar onto the first fraction bar. Copy the horizontal row lines from the first fraction bar onto the second fraction bar.

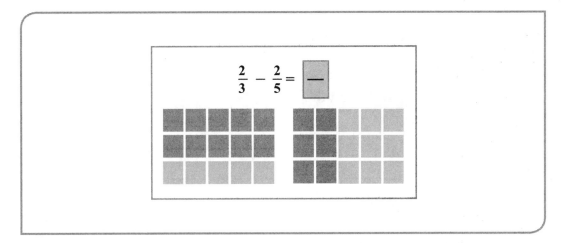

Step 4: Remove the smaller number of colored boxes from the fraction bar with the larger number of boxes. In this example, there are 10 colored boxes on the left and only 6 on the right. So subtract 6 boxes from the fraction bar on the left. Put an X over the now empty fraction bar to signify that we are done with it.

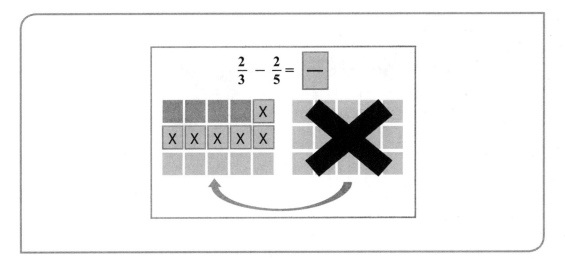

Step 5: Count the number of colored boxes that are left. This number is the numerator: 4.

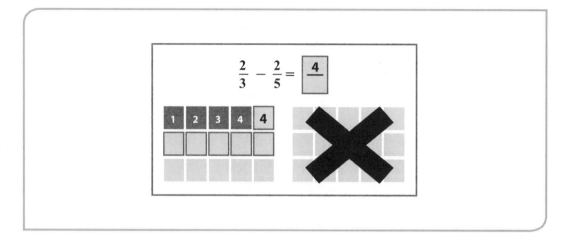

Step 6: Count the total number of boxes in the fraction bar. This is the common denominator: 15. Now write the equation, and reduce the answer if possible.

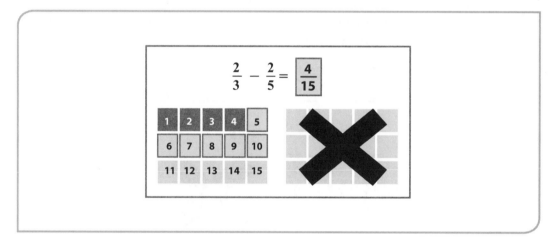

Answer: $\frac{2}{3} - \frac{2}{5} = \frac{4}{15}$

Activity 4

Multiplying Fractions

PROBLEM: $\frac{2}{3} \times \frac{1}{2} =$

Step 1: Create two fraction bars of the same size. Divide one into three vertical columns to represent the fraction with denominator 3. Divide the other into two horizontal rows to represent the fraction with denominator 2.

Step 2: Instead of coloring in both fraction bars, color the boxes in one and use a pattern on the boxes in the other to represent the numerators.

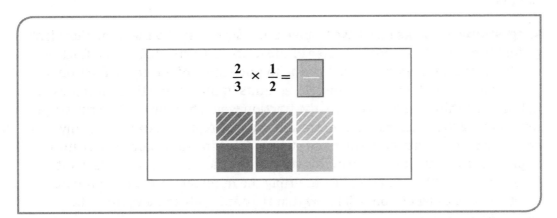

Step 3: Copy the horizontal lines and pattern from the second fraction bar onto the first. Tell students that they are laying the pattern from second bar over the first in the same layout as it is used in the second fraction bar. Now there is only one fraction bar.

Step 4: Count the total number of boxes in the fraction bar. This is the denominator: 6.

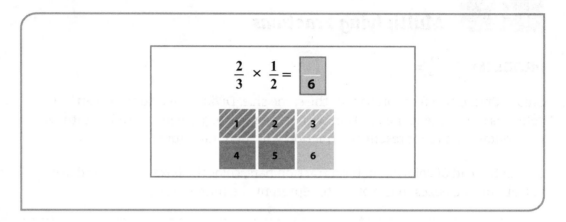

Step 5: Count the boxes that have both color and pattern. This number is the numerator: 2. Now write the equation, and reduce the answer if possible.

Answer: $\frac{2}{3} \times \frac{1}{2} = \frac{2}{6} = \frac{1}{3}$

Suggestions: Students may have trouble understanding why the numerator isn't 3, since the white lines cross all three of the units on the top half of the fraction bar. If so, remind them they aren't trying to calculate ½ of the entire fraction bar. The problem is ⅔ × ½ or ½ of ⅔. The colored units in the left fraction bar represent ⅔. If the last third was also colored, the fraction would be ³⁄₃ or 1. When the stripes representing the other fraction (½) are laid over the representation of ⅔, only two boxes are covered with both color and stripes. Give students many opportunities to practice this—their mastery of this concept is crucial. Once students get it, they might ask, "Why can't we just multiply the numerators and denominators together?" That's when you will know that they are ready to move from the concrete to the abstract.

Dividing Fractions

PROBLEM: $\frac{1}{4} \div \frac{2}{3} =$

Step 1: First review division terms with students. The first number is the dividend, and the second number is the divisor. The dividend (¼) is divided by the divisor (⅔). Create a fraction bar for the dividend and divide it into vertical columns to represent the denominator 4. Create another for the divisor and divide it into horizontal rows to represent the denominator 3.

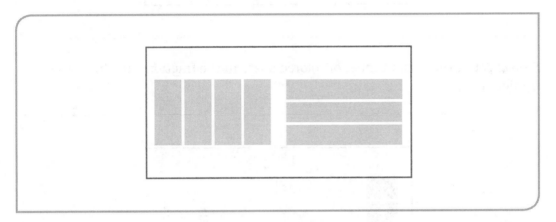

Step 2: Color the boxes in one bar, and use a pattern on the boxes in the other to represent the numerators.

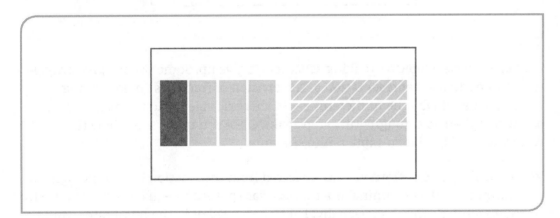

Step 3: Copy the horizontal lines and the pattern from the second fraction bar onto the first. Tell students that they are putting the second bar on top of the first, so now there is only one fraction bar.

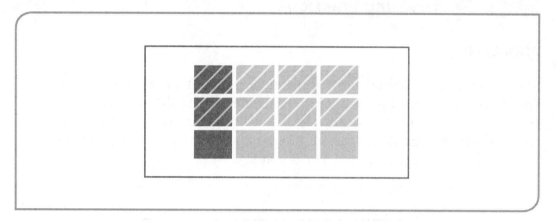

Step 4: Count the total number of colored boxes in the fraction bar. This is the numerator: 3.

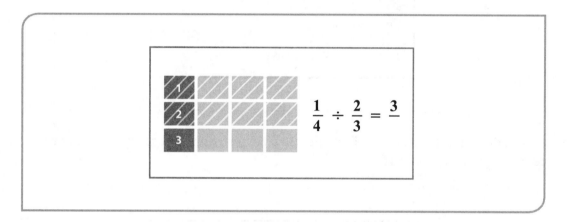

NOTE: If the last step isn't making sense, don't give up. Some people who struggle with this particular representation of the division of fractions get lost trying to make sense of why the numerator is 3 in this example. Read the explanation below, and then go through every step again. Once you understand how this works, you will be able to explain it to students.

When the lines and pattern were transferred from the fraction bar on the right to the one on the left, the original fraction (¼) was split into 3 equal units. That is why the numerator is 3. If ¼ had been divided by ⅓ instead of ⅔, the numerator would still be 3, because the divisor is splitting the dividend into thirds.

If you were dividing ⅗ by ¾, what do you think the numerator of the quotient would be? If you said 12, you would be correct. You would have 3 out of 5 sections of a dividend fraction bar shaded. Those 3 shaded sections would each have been divided by 4—the denominator of the divisor. The denominator tells how many units to divide the dividend into: 3 shaded areas, each divided into 4 parts = 12 units.

Step 5: Count the boxes that are covered with the pattern. This number is the denominator: 8. Now write the equation, and reduce the answer if possible.

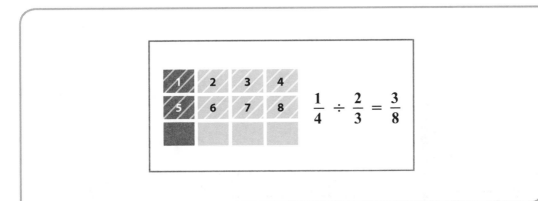

Answer: $\frac{1}{4} \div \frac{2}{3} = \frac{3}{8}$

 *Video Activity: Fraction Operations*

Decimals

In addition to teaching place value and relationships, whole number operations, and size/scale, base ten blocks are a useful tool for introducing students to decimals and operations with decimals. You can easily adapt the blocks by assigning new values to them. In fact, you can also adapt a place value mat and chips to work with decimals. Remind students to keep a list of new vocabulary in their journals.

WHOLE NUMBERS

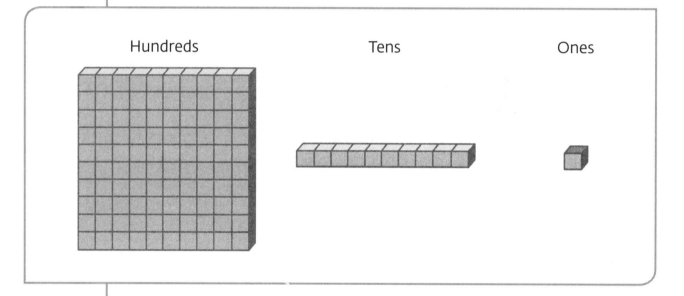

| Hundreds | Tens | Ones |

DECIMALS

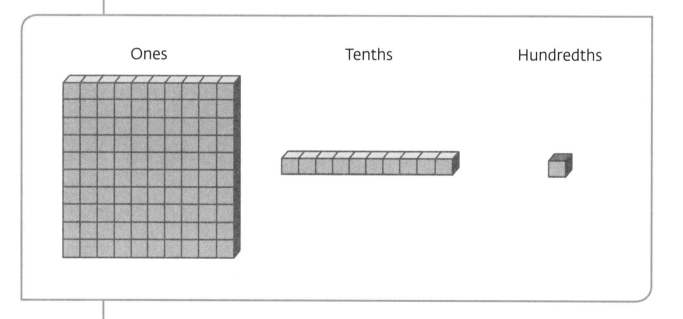

| Ones | Tenths | Hundredths |

Make sure students recognize that even though the values of the base ten blocks may be different from when they last used them, the relationships among the pieces remains the same. If you purchase base ten blocks, the blocks will be yellow, green, and blue. When working with whole numbers, the yellow cube represents 1, the green stick represents 10 times that, or 10, and the blue waffle represents 10 times that, or 100. Working with decimals, each piece has different values, which correspond to their move to the right along the place value continuum. Whether you purchase blocks or make your own, make sure students understand the value assigned to each block.

The best reason to use the base ten blocks in decimal instruction is the idea of scale. No matter what numbers you assign, the relationships stay the same: The stick (rectangle) is 10 times the small cube, and the waffle (large square) is 10 times the green stick. Here is one way to represent a decimal number using base ten blocks:

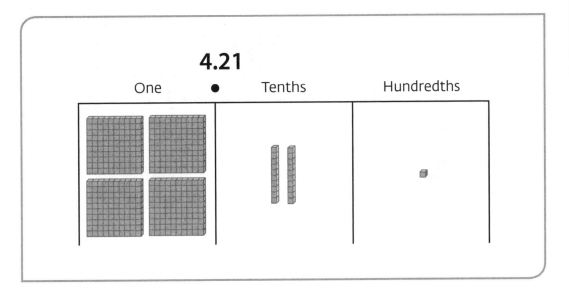

In this example, each waffle is 1; the rectangles are each one-tenth, and the cube is one-hundredth. According to the decimal number 4.21 (which is four and twenty-one hundredths), you have 4 ones, 2 tenths, and 1 hundredth.

When you reassign quantities to the blocks, allow students time to master the terms (tenths, hundredths, thousandths, etc.) and to understand the representations of decimal numbers before you move on to performing operations on decimals.

Addition and subtraction of decimals mirrors the same operations on whole numbers. You can use the examples in Chapter 8 with students. Just add decimals to each number.

The standard algorithm for multiplying decimals requires you to ignore the decimal point while performing the multiplication. Then you count the numbers after the decimal points in the multiplicands and—starting from the right side of the product—move that number of places to the left and place the decimal point. For more information on demonstrating this process, see the video activity on using base ten blocks.

Activity 6

Multiplying Decimals with Base Ten Blocks

PROBLEM: 3.3 × 2.1 =

Step 1: Represent the two multiplicands using blocks. Across the top row, place 3 waffles and 3 sticks to represent 3.3. Then down the left side of our array, place two waffles and one stick below it to represent 2.1. If it helps students conceptualize the process, place a decimal point between the large waffles (representing ones) and the tenths sticks.

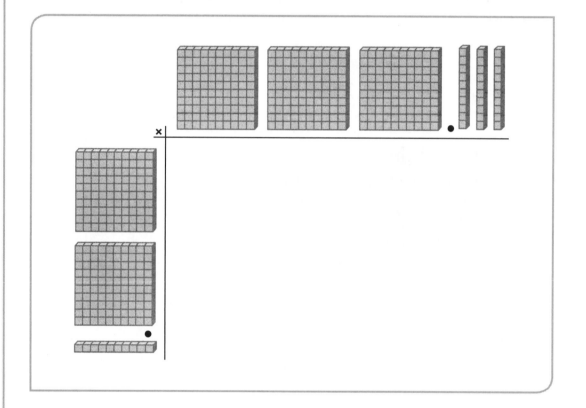

Step 2: To determine the product, multiply the first term in the vertical multiplicand by each term in the top row. Remembering our waffles have a value of 1, the first line of our multiplication array (inside the frame) is an identical copy of the vertical multiplicand.

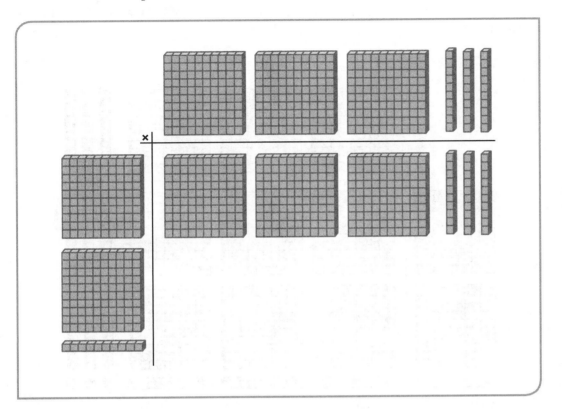

Step 3: The second row in the array is identical to the first because we multiplied the horizontal multiplicand by 1 again.

Step 4: The third row in the array is different because our multiplier is now 0.1. The first three terms in the third row are also 0.1, because they are being multiplied by 1. Ask students what terms (or blocks) they have that will fill in the bottom right corner in the array. It is not necessary that they know off the top of their heads that $0.1 \times 0.1 = 0.01$. The small squares each have a value of 0.01, and they complete the array.

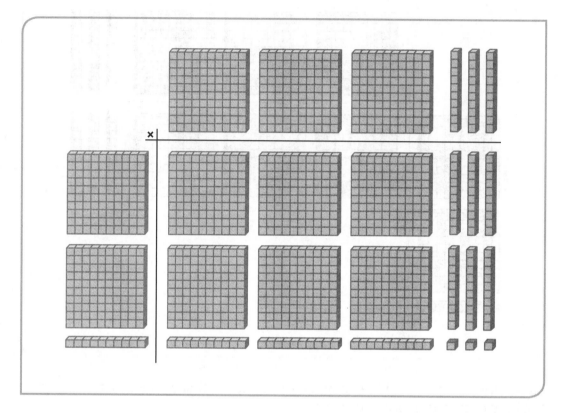

Step 5: To determine the product of 3.3×2.1, count up the total blocks under the frame:

- 6 ones = 6
- 9 tenths = 0.9
- 3 hundredths = 0.03

Answer: $3.3 \times 2.1 = 6.93$

Using base ten blocks like this will help your students conceptualize what is happening when they begin working with polynomials in algebra.

Factoring with Base Ten Blocks

PROBLEM: Find factors of −9.02

Look at an example of working backward using a finished rectangle to find the factors.

Look at this model using algebra tiles. Purchased tiles have a red, negative side. In this illustration, the gray tiles have negative values. What are the factors of this model?

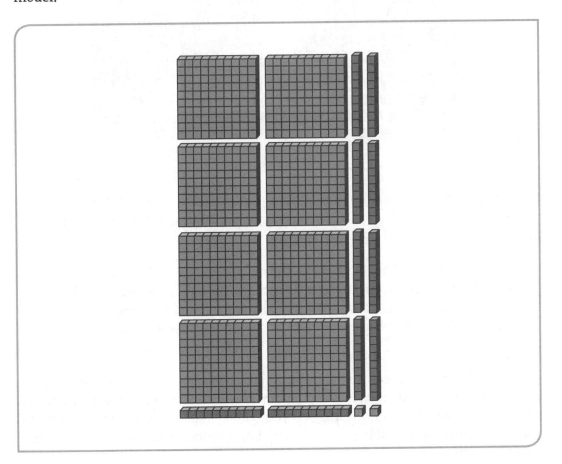

Read across the top: The value is −2.2.
Read down the left side: The value is −4.1.

Check the product: The entire model includes (negative) 8 ones, 10 tenths (or 1 one), and 2 hundredths. That totals −9.02.

−2.2 × −4.1 = 9.02
−2.2 × 4.1 = −9.02

One of the factors has to be negative, because the entire representation is negative. Does it make a difference which one is negative? No, it really doesn't right now. But in some problems it will matter.

Factoring is a form of division. What other factors are there? What happens if you divide this representation by 2?

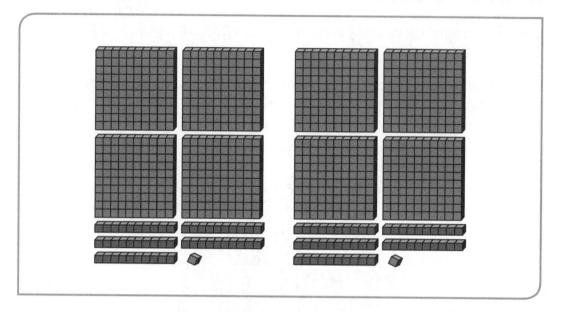

Now the entire representation is divided or partitioned into two groups of equal value. This is an example of partitive division, or sharing, because the whole amount has been divided into two equal parts.

You may be thinking, "Shouldn't one of the groups be positive?" So far, the whole has been divided into two equal parts, so each part is –4.51. The two factors of the original –9.02 are –4.51 and 2. The 2 is positive. If the factors were +4.51 and –4.51, our product would be between –16 and –25. (The answer would actually be –20.3401.)

Quotative division, or measurement division asks how many times a divisor will go into a dividend. For example, in the problem above, you could ask how many times –4.51 can go into –9.02, and the answer would be 2. Quotative division is helpful to use if you know the size of the part (divisor), but you don't know how many parts are in the whole group.

Unit Price

When discussing decimal operations, it's natural to use money as a relevant and nearly universal context. Using money makes the problems easier for students to relate to and understand. Calculating unit price is a skill that students can use every day.

It's easy to find examples of unit pricing, because most grocery stores, big box stores, and other retailers post it near the items for sale.

Here are some examples from a grocery store:

The tag on the left is for a 32-ounce bottle of a sports drink for $0.89. The picture on the right is for an eight-pack of the same drink, in 20 ounce bottles. It doesn't help to compare $0.89 to $5.99, since the amount of drink is different. But if you look in the lower left corner, you will see the unit price per fluid ounce. Explain to students that unit pricing is a way to compare costs when something is sold in different amounts.

Most people think that buying in bulk, or buying larger quantities of something, saves money. The real key to saving money is figuring out what the unit price is. In the section on fractions, I discussed unit fractions. Now I'll talk about unit decimals.

Explain to students that a unit price is actually a rate. In the example above, a 32-ounce sports drink is selling for 89 cents. To find the unit price, set up a proportion. The proportion compares the cost of the bottle of drink to the number of units it contains. The units will vary according to the item type. They may be servings, pieces, ounces, or in this case, fluid ounces.

Activity 8

Using Unit Price

PROBLEM: Which is a better value: a 32-oz.sports drink for 89 cents or eight 20-ounce bottles for $5.99?

Step 1: The price per unit equals the total cost of the item divided by the number of units it contains. Insert values for the sports drink into the equation:

$$\text{Unit price} = \frac{\text{cost}}{\text{number of units}}$$

$$\text{Unit price} = \frac{89 \text{ cents}}{32 \text{ ounces}}$$

Unit price = **2.78 cents per ounce**

Step 2: Notice that the proportion uses two different units of measure: cents and ounces. The unit price for the drink is calculated in cents per fluid ounce. Now compare this to the unit price of the multipack:

$$\text{Unit price} = \frac{\$5.99}{8 \times 20\text{-ounce bottles}}$$

$$\text{Unit price} = \frac{599 \text{ cents*}}{160 \text{ ounces}}$$

Unit price = **3.74 cents per ounce**

*Because the unit rate is cents per ounce, convert $5.99 to 599 cents.

With a unit price of 2.78 cents, the 32-ounce sports drink is the better value.

Recognizing the importance of unit rates can help students apply math in their daily lives, as well as save them money. In the two previous examples, the quantities were small, and both used cents per ounce. With other purchases, such as meat, the quantities may be larger, and the unit price may be in dollars per pound. Teaching students to either look for the unit pricing or figure it out themselves, can boost their confidence and their budgets.

Percents

Some students struggle to comprehend percents. This is why I recommend teaching fractions, decimals, and percent at the same time; they all address part/whole thinking. While you work on percent, it might help to have students keep a log of some of the more common fractions, decimals, and percents (remember equivalence of value) used in real situations. The circulars that come in the Sunday newspaper include great examples. This also is a good section to include in a math journal. Have students make their own lists of where they might use these skills and when percents could be important to them—maybe it's splitting restaurant checks or figuring tips, or wanting to get the best deal on purchases.

Activity 9

Which Is Better: 33% Off or 33% More?

PROBLEM: A bag of chicken wings costs $9.45 for 3 pounds. Which is a better value, 33% off the bag or 33% more chicken for $9.45?

Step 1: Use the percent equation to find the cost of the chicken at 33% off, if three pounds of chicken costs $9.45.

> **percent × whole = part**
>
> 33% × 9.45 = part
> .33 × 9.45 = part
> 3.12 = part
> $9.45 − $3.12 = $6.33

Step 2: Next, determine the unit rate. Divide the cost of the bag by the number of pounds. What is the unit price of the chicken at 33% off?

> $6.33 ÷ 3 = $2.11/pound

Step 3: Now, calculate the unit price of 33% more chicken.

> 33% is roughly ⅓, in fractional form, or .33 in decimal form. So, ⅓ more than 3 pounds, is one additional pound.

What is the unit price of 4 pounds of chicken?

$$\$9.45 \div 4 = \$2.36/\text{pound}$$

Step 4: Compare unit prices.

To find the better deal, compare the unit rates: 1 pound of chicken at 33% off to 1 pound of chicken at 33% more for the same price.

33% off: $2.11/lb
33% more chicken: $2.36/lb

At 33% off, the unit price is lower. So the answer is: **33% off**

Algebra and Functions

Variables, Expressions, and Equations

Math is a language. To be fluent in any language—to understand and be understood—requires speaking, listening, reading, and writing. It stands to reason, therefore, that students must be given ample opportunities to speak, hear, read, and write the language of mathematics. As instructors, we should encourage students to use the appropriate terms in class and small group discussions, and when describing reasoning and problem-solving processes.

Two of the most challenging topics for adult students are word problems and using variables. One way to help students prepare for both is to have students practice translating words into math expressions or equations and vice versa. On the following page is an example of a practice worksheet:

Math Expression or Equation	Word Problem
1. $3x + 2$	1. Three times a number plus 2.
2. $4 + 6(x - 2)^2$	2. **Four plus 6 times the squared quantity of the difference of a number minus 2.**
3. $9(x + 7)$	3. Nine times the quantity of an unknown number added to 7.
4. $(\frac{5}{8}x - 9) + 2 = 16$	4. **The difference of five-eighths of a number minus nine, added to two equals 16.**
5. $\frac{1}{2}(3x + 9)$	5. One-half the quantity of 3 times an unknown number plus 9.
6. $x(x^2 + 6x - 8)$	6. **An unknown number multiplied by the quantity of the same unknown number squared, added to six times the unknown number minus eight.**
7. $\frac{4(x - 9)}{2}$	7. 4 times the difference of a number minus nine, all over two.
8. $\sqrt[3]{27} + x(x - 2)$	8. **The cubed root of 27 plus a number times the quantity of the difference of the same unknown number minus two.**
9. $m = \frac{4 - 0}{0 - (-4)}$	9. The slope of a line passing through the point (−4, 0) and the point (0, 4).
10. $y = mx + b$	10. **The value of the point y is equal to the slope of the line times the x value of the point added to the y-intercept of the line.**

The first steps in solving a word problem are often the hardest: decoding it, determining what is known, and what is unknown, and figuring out what question you need to solve.

Activity 10

Decoding a Word Problem

Problem: A group of coworkers is going on a trip to see a baseball game. The employer is supplying transportation in the form of chartered 15-passenger vans. Nine employees, each with a spouse or date, and three children who require car seats will be traveling to the game. How many vans will be necessary to take everyone to the baseball game?

Student A Solution:

Student A adds up the people going to the game:

9 employees + 9 spouses/dates = 18 adults + 3 children = 21 people going to the ball game.

Answer: 21

Student A has correctly totaled the number of people, but she has not answered the question.

Student B Solution:

Student B totals the number of people going to the game: 21.

Each van seats only 15 people so he divides:

$$
\begin{array}{r}
1.4 \\
15\overline{)21.0} \\
\underline{15} \\
60 \\
\underline{60} \\
0
\end{array}
$$

Answer: 1.4 vans

Student B correctly divided the total number of people by the number that fit in each van. But he did not consider that fact that 1.4 vans is not a reasonable answer.

The division (using the standard algorithm once the dividend and divisor are chosen) was done correctly, but the question didn't just require accuracy in procedures. The problem required higher order thinking skills. Students need to think about how their answers relate to the question and assess them for reasonableness. This is why depending on procedure alone no longer adequately serves the needs of students.

Is there any other factor that could affect the number of people in each of the vans? The answer is yes: Who will be driving the vans? If the employees are responsible for driving, then all the information necessary to solve the problem is contained in the problem. If the employees are not responsible for driving, then the students must account for the drivers in their solutions.

Correct answer: 2 vans

Regardless of who drives, 1 van is not enough and 2 vans should accommodate everyone.

Using Algebra Tiles

Algebra tiles can help students (and teachers) understand what is happening in an algebra problem and solution. If you do not have the funds to purchase algebra tiles, refer to the template in Appendix C to make your own. If you make your own, use these standard colors:

- Large square: blue (positive)

- Long rectangle: green (positive)

- Small square: yellow (positive)

- Backs of all tiles: red (negative)

Put red on the back of each tile before you laminate it.

The following images illustrate the values of the Algebra Tiles. Look at the sides of the x^2 tile. The length of each is the same as the length of the green rectangle: x. Show students the tiles, and explain the value of each one.

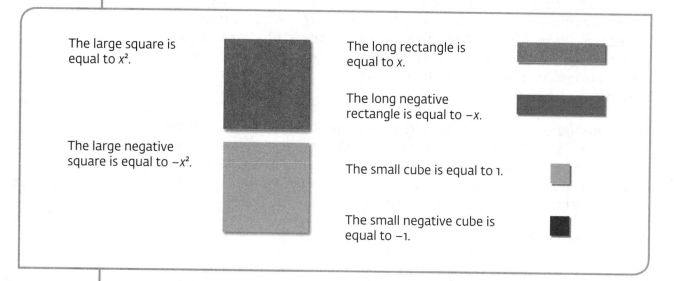

The large square is equal to x^2.

The large negative square is equal to $-x^2$.

The long rectangle is equal to x.

The long negative rectangle is equal to $-x$.

The small cube is equal to 1.

The small negative cube is equal to -1.

The bottom 3 images show how the value of x (the long rectangle) is the same length as each side of the x^2 tile. It also shows how, just as with integers, multiplying two positive or negative x tiles together will yield a positive x^2, while a negative x times a positive x will yield a negative x^2.

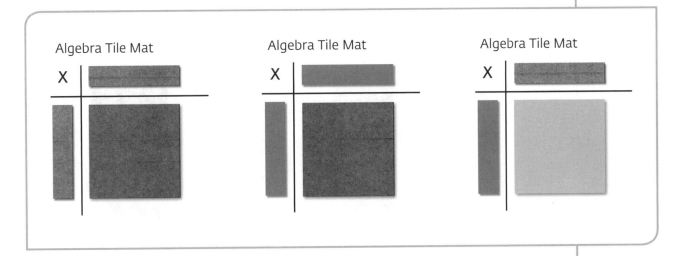

One of the most important uses for algebra tiles is to help students to group like terms and solve equations.

When students know the values of the tiles, it's time for them to start working with them. Make up a list of expressions and equations, and have students represent them at their desks. Extend the activity by having students perform computations (getting all terms on the same side of the equal sign, combine like terms, and remove any zero pairs that might exist).

Example Expressions to Represent:

$3x^2 - 2x - 9$

$2x^2 + 5x = x^2 - 8$

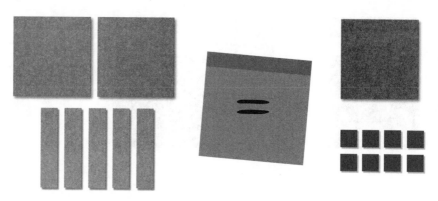

To solve this equation, move everything to one side of the equal sign, remembering to change signs when necessary. (This is assuming some basic numeracy skills, but if your students need more explanation or practice with this concept, do not skip it. Understanding why what is done to equations must be balanced is critically important!)

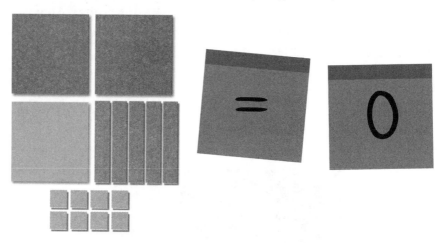

Finally, combine like terms, remove zero pairs.

$4x + 8 = x^2$

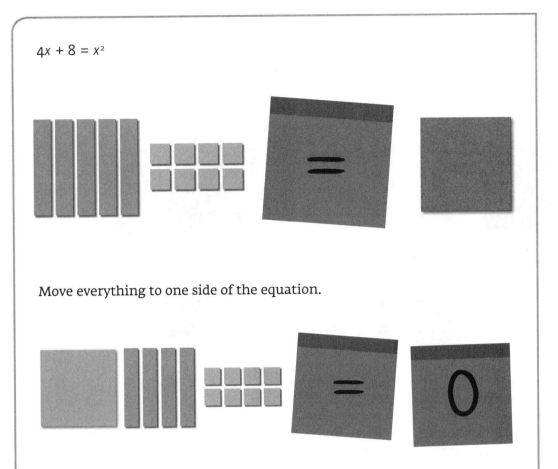

Move everything to one side of the equation.

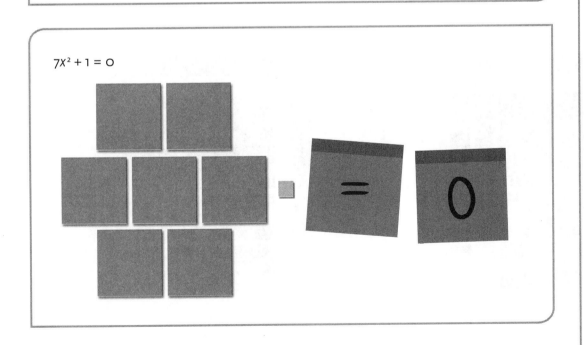

This representation is simplified as much as it can be.

$7x^2 + 1 = 0$

Next, ask students to represent different expressions. For example, "Use the tiles to represent the expression $x^2 + 6x + 8$."

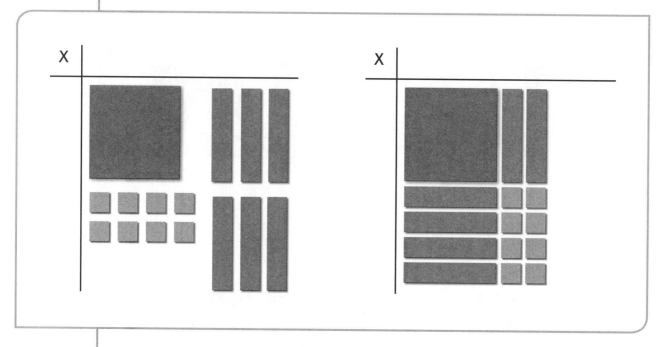

Both of these images represent the expression $x^2 + 6x + 8$, but the image on the right shows the tiles arranged in the most useful way. If you study the image on the right, you can see that the tiles form a complete rectangle with no gaps.

Now, look across the top of the rectangle on the right. What is the length of this side of the rectangle? $(x + 2)$ What is the length of the left side of the rectangle? $(x + 4)$.

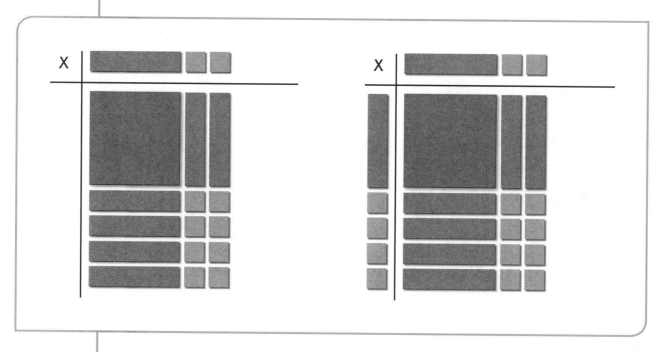

Algebra tiles are very versatile. In addition to adding or subtracting integers, or demonstrating how to combine like terms, it is also possible to factor polynomials with them. In fact, that is what you just did. Finding the lengths of the sides of a quadrilateral with algebra tiles is doing the same thing as finding the factors of a polynomial.

 *Video Activity: Using Algebra Tiles*

Adult Secondary Education Math

This book emphasizes crafting instruction that focuses on developing numeracy skills in order to provide students with the skills and understanding they need to do more sophisticated math. Numeracy skills involve more than the ability to apply an algorithm and make accurate computations. Numeracy requires mathematical thinking and persistence. Numeracy is problem solving. Numeracy means asking students to stretch their abilities and challenging them with rich assignments and activities.

According to the CCRS, there are five domains containing the major and supporting standards in what is considered the adult secondary education (ASE) levels. The CCRS uses level descriptors A through E, with A containing the beginning literacy standards, and E containing both low and high ASE content.

The five domains in the CCRS Level E are:

11. Number and Quantity
12. Algebra
13. Functions
14. Geometry
15. Interpreting Data

CCR Level E—Number and Quantity

There are three Level E standards in the Number and Quantity domain, and they deal with:

- rewriting expressions with radicals and rational exponents
- using units in multistep word problems
- choosing appropriate units when working problems to reflect the most accurate/appropriate solution

Before you address these topics with students, review the concept of rational numbers. A rational number is a number that can be expressed as a ratio or fraction in which both the numerator and denominator are whole numbers. The decimal equivalents of rational numbers terminate at some point, or they have a repeating pattern. Therefore, pi (π) is an irrational number, because its decimal equivalent neither repeats nor terminates.

It may also be helpful to review that a radical is any number or expression that contains a radical symbol ($\sqrt{}$). You might hear students refer to the radical symbol as the "square root" symbol, but that is not correct. The same symbol is used with square roots, cube roots, fourth roots, and more. So it is important to use the appropriate terms during instruction and discussions. You will also examine expressions with rational (fractional) exponents.

This is how to interpret a radical expression:

$$\text{index} \rightarrow \sqrt[3]{27} = 3 = 27^{1/3} \leftarrow \text{exponent}$$

base → (points to 27 in the exponent expression)

radicand ↑ (points to 27 under the radical)

An expression with a rational exponent could be, for example, $27^{1/3}$, which is interpreted as, "What number, multiplied by itself 3 times, equals 27?" Another way to look at this expression is "What is the cube root of 27?" You can factor 27 into $3 \times 3 \times 3$. Therefore, $27^{1/3} = 3$. The equivalent radical expression is $\sqrt[3]{27} = 3$. In general, if there is a radical symbol with no index, such as $\sqrt{25}$, the index is understood to be 2. This symbolizes the square root of a number.

Be sure that students have a firm grasp of these rules about exponents before you move on.

Rules of Exponents

	RULE	EXAMPLE
1.	An exponent is a notation that stands for repeated multiplication of a radicand by itself. The exponent tells us how many times to multiply the radicand by itself.	$4^3 = 4 \times 4 \times 4$
2.	When multiplying two like terms, both with exponents, add the exponents.	$(y^4)(y^7) = (yyyy)(yyyyyyy)$ $= yyyyyyyyyyy$ $= y^{11}$ $y^{(4+7)} = y^{11}$
3.	When adding, subtracting, or multiplying two terms, both with exponents, but with different bases, the exponents cannot be combined.	$(x^3)(y^5)$ This expression cannot be simplified.
4.	To calculate an exponential expression raised to a power, multiply the exponent and power together. The product becomes the new exponent.	$(x^3)^4 = (x^3)(x^3)(x^3)(x^3)$ $(x^3)^4 = x^{(3+3+3+3)}$ $(x^3)^4 = x^{(3 \times 4)}$ $(x^3)^4 = x^{12}$
5.	A negative exponent can be rewritten as the reciprocal of its base and its exponent. Then the exponent is positive.	$x^{-3} = \dfrac{1}{x^3}$
6.	Anything to the power of zero = 1, no matter the base.	$(x + y^3)^0 = 1$ This is well demonstrated using the law of negative exponents: $x^0 = x^{n-n}$ $= x^n \times x^{-n}$ $= \dfrac{x^n}{x^n}$ $= 1$

Remember that adding and subtracting fractions requires a common denominator. The same principle is at work in #3. Exponents cannot be combined unless they have common terms.

Use the rules of exponents to evaluate and/or simplify the following expressions.

1.	$(x + y^3)^3$	
2.	$81^{1/4}$	
3.	$(x^{3/8})(x^{1/4})$	
4.	$(x^3)^{-2}$	
5.	$((xy^3)(\sqrt[3]{(xy)^3})(xy^9)(xy))^0$	Answer: 1 (any term raised to the zero power is 1).

Algebra and Functions

Algebraic reasoning involves a great deal of problem solving and critical thinking and is a major component of all three high school equivalency exams. Students who plan to go to college need to apply the skills and concepts of algebra. Even those students who plan to go on to career and technical education (CTE) or directly into the workplace still need to be able to understand and apply mathematical concepts. Many career training programs—ranging from welding to health care—require the use of math.

The second and third domains in the CCRS are algebra and functions. The study of functions is complementary to algebra at the ASE level. Most of the algebra work at this level involves seeing structure in expressions, doing arithmetic with polynomials and rational expressions, and creating equations. Remember the Standards for Mathematical Practice (SMP) from Chapter 4? Discovering the concepts in this domain requires students to rely on more than one of the SMPs (especially #7 and #8).

Likewise, if you look at the standards that relate to functions at this level, they include interpreting functions and building functions. Within the strand of interpreting functions is examining linear and exponential functions. As you get started in the math for this domain/level, you may want to use a graphing calculator on a computer that is connected to a projector, so that you can display results to the class. One way to use tools like this is to have students predict what a change to the function will mean to the graph of its input or output. The website www.desmos.com offers a free, online graphing calculator. You can use the site to demonstrate graphing functions or when explaining input/output values.

Simplifying and Solving a Rational Equation

Here is an example of a higher level algebra activity that involves factoring.

Level E – Algebra

A.SSE.2 Use the structure of an expression to identify ways to rewrite it. For example, see $x^4 - y^4$ as $(x^2)^2 - (y^2)^2$, thus recognizing it as a difference of squares that can be factored as $(x^2 - y^2)(x^2 + y^2)$.

Problem: $\dfrac{7}{4x - 3} = \dfrac{5}{8x^2 + 2x - 6}$

Step 1: Examine the problem to see if or how it can be simplified. Remind students that this is similar to looking for the least common denominator before adding or subtracting two fractions. In this case, work backwards to factor one of the denominators so you can determine if that helps to solve the problem.

First look for a relationship between the denominators: Is $4x - 3$ a factor of $8x^2 + 2x - 6$? See if you can find a factor that when multiplied by $4x - 3$ equals $8x^2 + 2x - 6$.

Make the first term in the binomial factor $2x$, because $2x \times 4x = 8x^2$. You can think of the missing term as y: $(4x - 3)(2x + y) = 8x^2 + 2x - 6$

The last term in the polynomial is -6, so try substituting positive 2 for y: $2 \times -3 = -6$

$(4x - 3)(2x + 2) \overset{?}{=} 8x^2 + 2x - 6$

Step 2: To check the factors, multiply them using the FOIL method:

F: First terms in each binomial	$4x \times 2x = 8x^2$
O: Outer terms	$4x \times 2 = 8x$
I: Inner terms	$-3 \times 2x = -6x$
L: Last terms	$-3 \times 2 = -6$

Then, combine like terms:

$(4x - 3)(2x + 2) = 8x^2 + 8x - 6x - 6$
$(4x - 3)(2x + 2) = 8x^2 + 2x - 6$

Step 3: Now return to the original problem, and replace $8x^2 + 2x - 6$ with its factors: $(4x - 3)(2x + 2)$

$$\dfrac{7}{4x - 3} = \dfrac{5}{(4x - 3)(2x + 2)}$$

You can simplify the fractions by multiplying both sides of the equation by $(4x - 3)$:

$$(4x - 3)\,\frac{7}{4x - 3} = \frac{5}{(4x - 3)(2x + 2)}\,(4x - 3)$$

Don't say, "The terms cancel each other out." Choose your words carefully, so you don't confuse your students. Explain that when you multiply by $4x - 3$, you actually multiply by the fraction $\frac{4x - 3}{1}$. Because $\frac{4x - 3}{4x - 3}$ equals 1, you can cross it out.

$$\cancel{(4x - 3)}\,\frac{7}{\cancel{4x - 3}} = \frac{5}{\cancel{(4x - 3)}(2x + 2)}\,\cancel{(4x - 3)}$$

Step 4: Rewrite the simplified problem and solve for x.

$$7 = \frac{5}{(2x + 2)}$$

Cross multiply.

$$\frac{7}{1} = \frac{5}{(2x + 2)}$$

$$7(2x + 2) = 5(1)$$

$$14x + 14 = 5$$

Subtract 14 from both sides to isolate the variable:

$$14x = -9$$

Divide both sides by 14:

$$x = \frac{-9}{14}$$

Note that we get only one solution using this solution path.

Step 5: Another solution path for this problem is to cross multiply and then factor, as if you were using algebra tiles.

$$\frac{7}{4x - 3} = \frac{5}{8x^2 + 2x - 6}$$

$$7(8x^2 + 2x - 6) = 5(4x - 3)$$

$$56x^2 + 14x - 42 = 20x - 15$$

Next, get all terms on one side of the equal sign, combine like terms, and set the entire equation equal to zero:

$$56x^2 - 6x - 27 = 0$$

Step 6: Factor this quadratic equation to find the solution.
I recognize this is challenging, but remember, you already know one of its factors!

Try the method you used to factor $8x^2 + 2x - 6$. You know one of the factors is $4x - 3$, so first figure out how many times 4 goes into 56.

$$56x^2 - 6x - 27 = 0$$
$$(4x - 3)(14x + y)$$

To find y, look for a number that, when multiplied by the other factor, equals -27. If the other factor is $4x - 3$, and $3y = -27$, then $y = 9$.

The factors for this quadratic equation are:

$$(4x - 3)(14x + 9) = 0$$

Step 7: Set both factors equal to zero: $4x - 3 = 0$ $\qquad$ $14x + 9 = 0$

Isolate x on one side of the equation by adding the inverse of each integer to both sides:

$$
\begin{array}{c}
4x - 3 = 0 \\
+3 \quad +3 \\
4x = 3 \\
x = \frac{3}{4}
\end{array}
\qquad\qquad
\begin{array}{c}
14x + 9 = 0 \\
-9 \quad -9 \\
14x = -9 \\
x = \frac{-9}{14}
\end{array}
$$

Step 8: Check your answers. To check if one, both, or neither of the solutions are correct, substitute each into the original equation. First try $x = \frac{3}{4}$, because we already know $\frac{-9}{14}$ is a correct solution.

$$\frac{7}{4x - 3} = \frac{5}{8x^2 + 2x - 6}$$

Tip: Use the cross product from Step 5, and set it equal to zero. This makes the computation easier.

$$56x^2 - 6x - 27 = 0$$

$$56\left(\tfrac{3}{4}\right)^2 - 6\left(\tfrac{3}{4}\right) - 27 = 0$$

$$56\left(\tfrac{9}{16}\right) - \tfrac{18}{4} - 27 = 0$$

$$\frac{504}{16} - \frac{72}{16} - 27 = 0$$

$$\frac{432}{16} - 27 = 0$$

$$27 - 27 = 0$$

So, both of our answers are solutions to this quadratic equation. $x = \frac{-9}{14}$ and $x = \frac{3}{4}$

Activity 12

Designing Flower Beds

The goal of this activity is to create opportunities for reasoning, recognizing patterns, and structuring problem solving.

Problem: You have a summer job with the Parks and Recreation Department. You show up for work one morning, and your supervisor tells you to design a new flower bed for a local park. You must design the flower bed and order the materials. You need to outline the bed with paver tiles, and you need to order exactly the right number of pavers because money is tight. She tells you the pavers that you will use measure 1 foot square and are $7 each. There is a minimum order of $75 for the pavers. Your flower bed can be any shape, but your design should allow for maximum planting area. Your total budget cannot exceed $325 for the tiles. There may be money to enlarge the flower bed in the future, so think about whether the design would be easy to modify.

NOTE: I suggest using rolls of 1-inch graph paper with 1 square inch foam tiles for activities like these. These materials are usually available at school supply stores or online, or you can cut square tiles out of construction paper.

As you begin the activity, ask students to think about what they would do first. What questions do they want to ask? Here are some questions that may come up before or during the activity:

1. What is the maximum number of tiles you can use and still stay within the allowed budget? (325 ÷ 7 = 46)

2. What is the largest flower bed that can be created with a border of these tiles without going over budget?

3. What shape flower bed will give you the largest area for flowers?

4. What design might be easiest to modify if improvements are made in the future?

5. What is the smallest flower bed you can create using the pavers?

6. How many tiles would you have to add to the border in order to add one more square foot of planting area to the horizontal flower bed below?

7. Look at the horizontal flower bed on the following page. How would the number of tiles you need change if you made the planting area 2 feet wide, instead of 1 foot wide, as in the diagram below?

8. What algebraic expression could you use to find the number of tiles needed to border a flower bed with n number of square feet for planting?

Possible Solution Path: One long horizontal flower bed

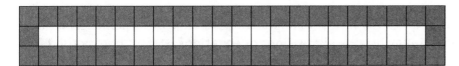

In this rectangular design, there are 46 tiles. The width of the area for planting flowers is only one tile or one foot, so the total area for planting flowers is 20 square feet.

The easiest way to think about modifying the design in the future is to think about what will stay constant and what will change. For example, consider how the number of tiles is affected when you add another row of flowers or if you add to the length of the flower bed. If you are building a rectangular flower bed, you can look at this two different ways: either the number of tiles on each end stays constant (vertical sides), or the number of tiles on the long (horizontal) sides stays constant. Let's look first at a design where the number of tiles on the ends stays constant. For each square foot of planting area you add, the total number of tiles will increase by two—one tile on the top and one on the bottom.

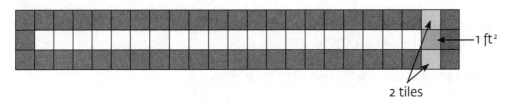

When one square foot is added for planting, two tiles are added to the border. The three tiles on each end remain constant, for a total of six. You can represent this algebraically:

$$t \text{ (number of tiles)} = 2n \text{ (number of square feet for planting)} + 6$$

So, if you build a simple rectangular design such as this, with tiles that cost $7 each, you could purchase 46 tiles with your $325 budget. But is this the best use of the budget? The area of the flower bed seems quite small compared to the number tiles used.

A great extension exercise for this type of problem is to challenge students to find the arrangement with the largest planting area or find the best shape for maximizing the planting area. It might be worth pointing out to students that it's possible to create flower beds using fewer tiles that have a much larger area. For example, the flower bed below uses only 38 tiles, and has 52 square feet of planting area.

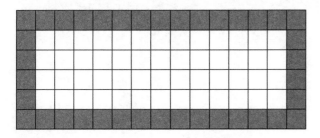

Polynomials

Polynomials can be used in a variety of applications. Sometimes operations with polynomials result in fairly large expressions or equations. Here is one method for multiplying polynomials that will help students keep track of all the terms of a polynomial product.

Activity 13

Multiplying Polynomials Using the Box Method

Problem: Multiply $(4x^3 - 4x - 7)(3x^2 + 4x + 3)$

Step 1: Rewrite any polynomial, replacing subtraction with addition of a negative term.

$$(4x^3 + (-4x) + (-7))(3x^2 + 4x + 3)$$

Step 2: Insert a place holder variable to hold a place for any terms not included. In this case, there is an x and an x^3, but no x^2.

Place holder variable: $(4x^3 + 0x^2 + (-4x) + (-7))(3x^2 + 4x + 3)$

Step 3: Construct a box with the same number of cells as there are terms in each polynomial.

	$4x^3$ +	$0x^2$ +	$-4x$ +	-7
$3x^2$ +				
$4x$ +				
3				

Step 4: Draw diagonal lines in the boxes and multiply.

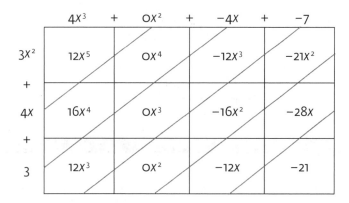

Step 5: Finish by combining like terms (look along the diagonal lines). Then simplify.

$$(4x^3 - 4x - 7)(3x^2 + 4x + 3) = 12x^5 + 16x^4 - 12x^3 + 12x^3 - 21x^2 - 16x^2 - 28x - 12x - 21$$
$$= 12x^5 + 16x^4 - 37x^2 - 40x - 21$$

This method—simply based on a form of place value—makes managing all the terms in polynomial multiplication much more manageable. This might remind you of lattice multiplication.

Functions

Helping students understand functions begins with helping them conceptualize what a function is. I find the easiest way to begin this is to use a visual display. I made these graphs with the free graphing calculator on Desmos.com.

Function	Linear
Input : Output	1:1
Display	straight line
Example	$f(x) = \frac{2}{3}x + 4$
Graph	
Explore	• What does the shape of the graph tell you about the relationship between input and output values? **(Each input has one and only one output value.)** • What does the graph tell you about the function being modeled? **(The line slopes upward from left to right, so it is positive.)** • What type of situation could be modeled with this function graph? **(Something that steadily increases, such as, payment for hours worked, etc.)** • What would happen if: ▷ The y-intercept changed? **(The point at which the line crosses the y-axis would be moved up or down, depending on its value.)** ▷ The slope of the line changed? **(The line will be come more steep or more flat depending on the value.)** ▷ The input value changed (e.g., we use 9 for the *x* value)? **(Nothing will happen to the graph, because if we input 9, the output value is 10. The point (9, 10) is on the graph of the line.)**

Function	Quadratic
Input : Output	1:2
Display	parabola
Example	$f(x) = 3x^2 + 4$
Graph	

Function	Exponential
Input : Output	1:1
Display	Approach an *asymptote*— a point on the *x* axis it approaches but never quite reaches
Example	$f(x) = 3^x + 2$
Graph	

Other functions include interest accrued on loans or interest earned on other accounts, such as savings accounts, certificates of deposit, etc. The following activity uses an equation to calculate interest, that is also considered a function of time and interest, applied to the amount of money you borrow.

Activity 14

Buying a Car

Problem: You need to purchase a car. You find a used car that you like, but you are in a hurry to get it. You can't pay cash for the car, so you need to finance it. Now comes the hard part: negotiating a car loan. To do that, you have to know how interest works.

The bank will pay the dealership the cost of the car, and you repay the money to the bank, along with a fee you pay to borrow the money (interest). The amount of interest you pay depends on the amount borrowed and the length of time it takes you to pay back the loan. The better your credit rating is, the lower the interest rate is that you will pay. If you have bad credit, you will be charged a higher rate in order to borrow the money.

Here is a simplified version of a loan interest formula:

$$y = \text{Principal} \frac{(1 + (\text{Interest Rate})(\text{Time in Years}))}{12\ (\text{Time in Years})}$$

Where y = the payment, and principal = the amount of money being borrowed.

The interest rate and the length of the loan are variables. Use this formula to calculate payments, and complete the function table. Round your calculations to the nearest whole dollar. For this activity, the price of the car is $6,500 and the interest rate is 7.5%. (You can devise new activities by having students look up prices and interest rates in a newspaper or online.)

For 24 months:	For 36 months:	For 48 months:
$y = 6500 \frac{(1 + (0.075)(2))}{12 \times 2}$	$y = 6500 \frac{(1 + (0.075)(3))}{12 \times 3}$	$y = 6500 \frac{(1 + (0.075)(4))}{12 \times 4}$
$y = 6500 \frac{(1 + 0.15)}{24}$	$y = 6500 \frac{(1 + 0.225)}{36}$	$y = 6500 \frac{(1 + 0.3)}{48}$
$y = 6500 \frac{1.15}{24}$	$y = 6500 \frac{1.225}{36}$	$y = 6500 \frac{1.3}{48}$
$y = 6500(0.0479)$	$y = 6500(0.03403)$	$y = 6500(0.02708)$
$y = 311.35$	$y = 221.20$	$y = 176.02$

x	y
24	$y_1 = 311.35$
36	$y_2 = 221.20$
48	$y_3 = 176.02$

You may want to have a class discussion about monthly payment vs. total price. Is the monthly payment or total price more important to students for a big ticket item, such as a car? Explain that the lowest monthly payment often adds up to the highest total price. Have students decide whether they would choose the 2, 3, or 4 year loan. Then ask them to multiply the number of payments (the x value) by the amount of each payment (the y value) to calculate the total amount that they would have to pay for the car to satisfy the loan.

For 24 months:	For 36 months:	For 48 months:
Total cost = $24 \times y_1$	Total cost = $36 \times y_1$	Total cost = $48 \times y_1$
Total cost = 24×311	Total cost = 36×221	Total cost = 48×176
Total cost = **$7,464**	Total cost = **$7,956**	Total cost = **$8,448**

Other Functions

In addition to linear functions, there are also exponential and quadratic functions.

When will students encounter exponential and quadratic equations? Aside from a high school equivalency test-prep math class, they are likely to encounter these functions in financial circumstances—especially when calculating simple or compound interest.

The quadratic formula: $x = \dfrac{-b \pm \sqrt{b^2 - 4ac}}{2a}$

The quadratic equation: $ax^2 + bx + c = 0$

You used algebra tiles to demonstrate multiplication of polynomials, and you can do the opposite operation—factoring quadratics—with them as well. Some quadratics are easy to factor. For example:

$x^2 + 5x + 6$ can be factored into $(x + 3)(x + 2)$.
(Look to find two factors that when added together equal 5, and when multiplied, equal 6.)

Activity 15

Factoring a Quadratic Equation with Algebra Tiles

Problem: Solve the equation by factoring: $6x^2 - x - 2 = 0$

There are several ways you can attempt to solve this. One is to attempt to factor it just by looking for two factors that add up to −1 and when multiplied equal −2. The coefficient of 6 makes this challenging. Here's a way to factor the quadratic equation using algebra tiles.

Step 1: Represent the expression using algebra tiles:

Remember:

- Large square = x^2
- Large negative square = $-x^2$
- Long rectangle = x
- Long negative rectangle = $-x$
- Small cube = 1
- Small negative cube = -1

Step 2: Add zero pairs to fill in the rectangle. To begin, move the (negative) one tiles out of the way momentarily. Because you already have a negative x on the right side, place the other negative tiles there as well:

Here are 3 zero pairs.

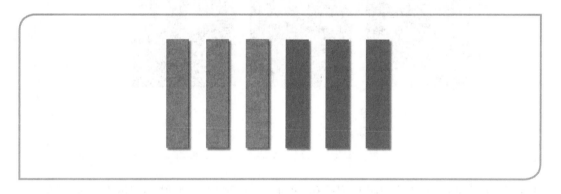

Try arranging the 3 zero pairs to see if you can complete the rectangle:

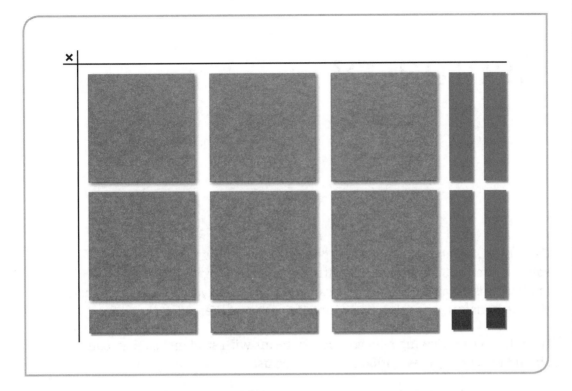

After you insert the 2 negative one tiles, you have a complete rectangle.

Step 3: Read the factors. The first factor, reading across the top, is $3x - 2$. The second factor, reading down the left hand side, is $2x + 1$.

Step 4: Set each factor equal to zero, and perform the calculations necessary to isolate the variable.

$$3x - 2 = 0 \qquad\qquad 2x + 1 = 0$$
$$3x = 2 \qquad\qquad\quad 2x = -1$$
$$x = \frac{2}{3} \qquad\qquad\quad x = -\frac{1}{2}$$

The solutions are: $x = \frac{2}{3}$ and $x = -1$

This is a great method for helping students see what's happening with the quadratic formula and quadratic equations. Starting with algebra tiles—even if your students are in ASE classes and have been doing abstract math—when teaching a new concept can be an ideal strategy. If students have had success with algebra tiles in the past, they are familiar with the values and the operations that can be modeled using them. You can build on that success by allowing them to continue to work with them to build conceptual understanding of these complex operations.

For more information and lessons on algebra and functons, see *Math Sense 3: Focus on Analysis*, at newreaderspress.com.

Geometry

Geometry is all about shapes, their properties, and how to use them to solve problems. Geometry, along with algebra, is easy to relate to real-world applications. As an instructor, you can craft some really rich lessons and activities that demonstrate the relevance of geometry to real life.

In addition to reviewing new vocabulary terms with students as they come, you should review any new symbols that will be used.

GEOMETRIC SYMBOLS

△   triangle

≅   congruent/congruency

≈   similarity/similar to

∠   angle (usually followed by one or 3 letters)

≠   "does not equal" or "is not equal to"

⌐   90° angle

To have success in geometry, students will need to classify shapes and figures based on unique features and properties. They will also need to use their knowledge of shapes to decompose and recompose them to solve problems. There is more vocabulary in geometry than in any other math topic, and it will take great organizational skills for students to keep track of the properties of each shape. If you have not already introduced math journals and/or math word walls to your class, this would be a great time for students to use them.

In addition to knowing the properties of common geometric shapes, students will need to understand the process for calculating area and perimeter. To ensure that students understand why and how these processes work, we can break down the geometric properties—essentially using the same conceptual framework we used to decompose and regroup numbers.

For the high school equivalency tests, students do not have to memorize all the formulas. They will be able to refer to math formula sheets, but the sheets do not necessarily include every formula required on the tests (see Appendix F). If students are studying for one of the tests, show them the appropriate formula sheet and have them use it in class for practice. Here are the formulas for finding the area of some common shapes:

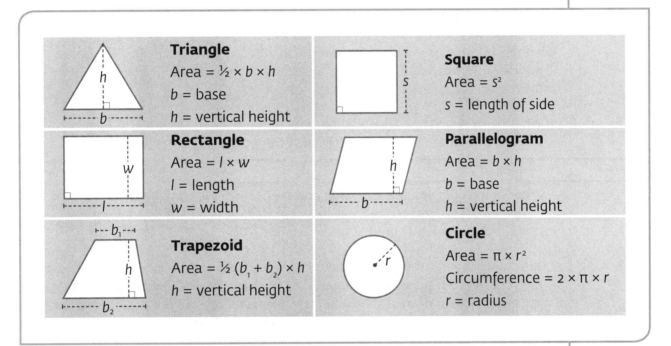

To ensure that students thoroughly understand geometric concepts, use activities that require them to think strategically and reason out a solution path, rather than those that simply require computation.

Composite Shapes

It is possible to find the area of a 2-dimensional shape by reasoning about its components rather than relying on a less familiar formula. Some shapes can be broken down into simple parts that students can more easily relate to. For example, consider the following approach to working with trapezoids.

Trapezoid: A trapezoid is a quadrilateral with only one pair of parallel lines. What makes this shape different from a parallelogram is that both sets of opposite lines in a parallelogram are parallel to each other. Let's take a look at a trapezoid and compare it to a parallelogram.

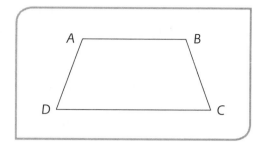

Trapezoid *ABCD* has one set of parallel lines: *AB* and *CD*. If these lines were extended infinitely out from the diagram, they would never meet.

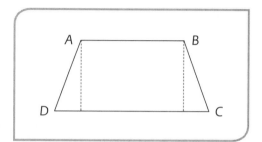

I mentioned decomposing and regrouping shapes before, and here is why I think it's a valuable skill to use when computing area and perimeter: We can reason out why and how the formulas for area and perimeter work.

What happens to the shape if you drop a straight line from *A* and *B* respectively to the base or segment *CD*?

What shape(s) do you see now? Add points E and F to the diagram to make it easy for students to talk about the shapes.

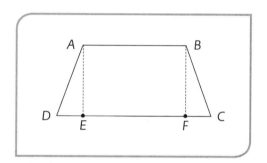

Ask students what they notice about the two triangles that have been formed on each end of the trapezoid? They appear to the be same size, don't they? Explain: If you are precise in drawing the lines, the two triangles will not just be equal in size, they will be congruent. There is a fixed relationship between Δ*ADE* and Δ*BCF*. Each corresponding part of the two triangles is equal, from the measure of the angles to the lengths of the sides.

What does that mean? It means you can't just decide to change the order of the letters representing one of the triangles without doing the same to the other triangle. We "name" a triangle by using the letters that correspond to each of its three vertices. With congruent triangles, we have to respect the relationship the two triangles share. We show congruency between two or more shapes with the ≅ symbol. We say, "Δ*ADE* is congruent to Δ*BCF*." Δ*ADE* ≅ Δ*BCF*

Now when you look at the trapezoid, what do you see? (a quadrilateral and two congruent triangles) What do you know about the quadrilateral bound by the two triangles? The new line segments *AE* and *BF* are perpendicular to *CD*. What does that tell you about the measure of ∠*AED* (or ∠*BFC*)? If you said it measures 90°, you would be correct. What does that tell you about the measure of ∠*AEC* (or ∠*BFD*)?

How does this new knowledge help you to compute the area of the trapezoid? Rather than trying to find the area of an unfamiliar shape, suggest that students decompose figures—wherever possible—into shapes that are easier to work with. The following activity makes use of many of these shapes and strategies.

Activity 16

Designing a Tile Floor

You work for a tile company and have been given the task of designing a floor with tiles shaped like isosceles triangles and squares in gray, dark gray, green, and light green. The base of each triangle tile is 12", and each one costs $8, no matter the color. The square tiles are each 1 square foot, and they vary in price. Light green squares are $10 each, dark green squares are $9, and gray squares are $7.50. If the floor of the room to be tiled is 23 feet by 14 feet, create a floor design and calculate the cost of the tiles.

There are many possible configurations, but here is one example:

Step 1: Determine if there is a maximum total price.

Step 2: Create a scaled design using graph paper, with one square equal to one foot.

Step 3: Total up the cost of the tiles, based on the design you create.

Step 4: Present (on paper) a professional quote for the cost of the tiles.

Step 5: Have students post their designs and quotes in the classroom. Have students circulate to look at each other's designs. Have students check the accuracy of at least two other quotes.

Step 6: Process the activity as a whole class discussion.

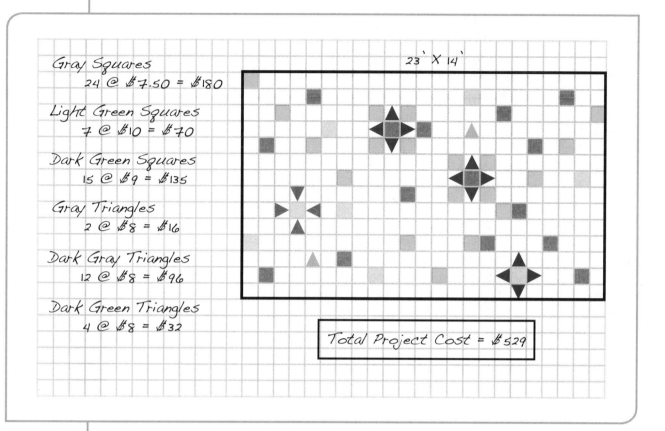

Gray Squares
24 @ $7.50 = $180

Light Green Squares
7 @ $10 = $70

Dark Green Squares
15 @ $9 = $135

Gray Triangles
2 @ $8 = $16

Dark Gray Triangles
12 @ $8 = $96

Dark Green Triangles
4 @ $8 = $32

23' X 14'

Total Project Cost = $529

This project assumed no cost for the plain colored tiles (but explore what it would do to the design/cost if there was a cost for those non-colored tiles as well. Discuss what would happen if the client wanted a repeated pattern vs. a random pattern, as in the picture.

 **Video Activity: Geometry with Pattern Blocks**

For more information and lessons on geometry, see *Math Sense 2: Focus on Problem Solving*, at newreaderspress.com.

Data and Statistics

Data is everywhere. Everyday decisions are based on the data you receive. What is the temperature? What is the chance that it will rain this week? It's important to know how to use and interpret data that you encounter every day.

I live right in the middle of Illinois—two hours from Chicago, and 3 hours from St. Louis, Missouri. I am interested in weather data, and I pay attention to information about climate change. Many people can read a list of data and interpret meaning from it. I am not one of those people. As a visual learner, I find it easier to understand a chart or graph than to read a list of numbers.

On the following page is a table showing the high temperature readings in my hometown for each day in January, over five years.

DAILY TEMPERATURE READINGS IN JANUARY 2010–2014 IN BLOOMINGTON, IL

Date	2010	2011	2012	2013	2014
1	30	55	45	34	25
2	10	24	37	21	21
3	6	26	24	24	15
4	9	37	28	30	21
5	12	25	45	30	35
6	9	36	52	34	19
7	14	30	55	28	−9
8	17	26	44	36	9
9	18	19	37	41	14
10	10	24	45	48	30
11	21	30	53	53	37
12	19	25	51	55	37
13	28	26	21	48	44
14	35	15	17	21	46
15	39	28	21	23	36
16	28	26	33	27	33
17	28	23	52	34	32
18	30	35	42	33	12
19	30	33	26	42	23
20	26	21	26	50	37
21	30	19	19	21	37
22	32	10	24	10	7
23	34	19	46	14	18
24	43	27	43	30	5
25	44	27	34	21	33
26	30	28	32	34	32
27	18	24	34	30	39
28	25	28	39	45	5
29	15	28	39	45	10
30	16	35	32	62	28
31	24	33	55	48	34

I entered the data I gathered into an Excel spreadsheet, because Excel makes working with data and charts easy. Using a spreadsheet allows me to not only interpret the data, but to also choose how to represent that data. I can choose different types of charts or graphs, such as bar graphs, a line graphs, or pie charts.

Here is a line graph of the temperature data I entered. The questions below the graph can be answered by looking at the raw data, the graph, or both. Keep in mind that in some instances—such as on HSE tests—students may not be able to refer to the raw data. They may have to make inferences from the graph or chart given.

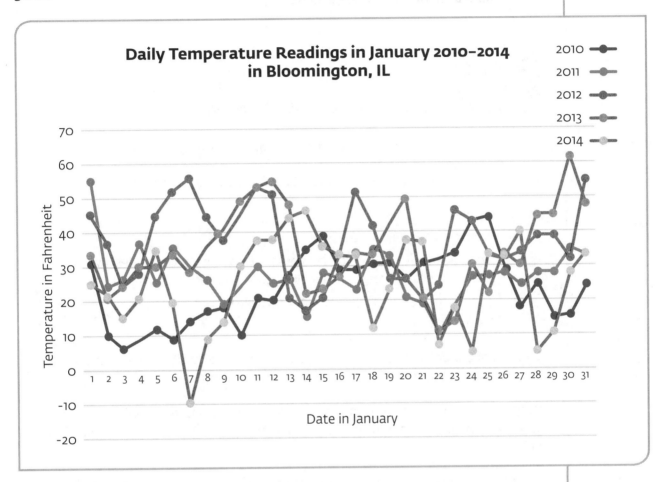

Questions to interpret data:

- How many variables are being shown? (**two: temperature in Fahrenheit and date**)

- What is the range of the temperature values? (**–9.9 to 62.1, but you can also ask students to calculate the mean, median, and mode, and to use the data to determine the coldest or warmest January during the five-year period**)

- Which of the five years had the coldest January? (**2014: according to the graph, because it shows the five lowest temperatures on the graph**)

- Can you make an argument either supporting or denying the existence of global warming based on the data and the graph? (**Allow students to take sides, and conduct some more research to gather data to substantiate or refute the claim of global warming.**)

- Although its data is not represented, the winter temperatures in January of 2016 were uncharacteristically mild. How might adding this data affect the graph overall? (**Cannot be determined, but can create some great discussion**)

- Does precipitation affect temperature? (**Cannot be determined from this graph, but can be an extension for some students to pursue finding the data and making an argument**)

Activity 17

Gathering Data with Beans

Problem: Gather and manipulate data to simulate inherited traits

Pre-teaching topic: Discuss population, natural selection, inherited traits

Materials needed:
- Plastic knives, forks, and spoons
- Dried beans of various colors/types (I buy a bag that includes a variety of beans. You can change the labels to fit the types of beans you use.)
- A room to conduct the experiment/activity
- A stop-watch or timer

Concept of the activity: to demonstrate how inherited traits can determine survival or extinction, through data gathering of simulated generations

Step 1: Divide class into four groups. (It's okay for a group to include only one student.) Assign one tool to each group: plastic spoons, plastic forks, plastic knives, or hands.

Step 2: Pour the beans on the floor or onto a large table. Set the timer, and give students three minutes to pick up as many beans as possible using only the tool they were assigned.

Step 3: When time is up, have students collect and record their data (the number of each type of bean picked up).

Step 4: Continue timed rounds (generations) of picking up beans until all students have had a chance.

Step 5: Have students total the data for each type of bean collected.

Step 6: Genetics: Tell students that each tool represents a genetic trait. Using an average 25% survival rate, have them determine which traits will not survive. For example, if a total of 250 yellow beans were collected in the first round, the 25% survival rate would mean each method of gathering would have to have collected at least 63 beans. If any one of the gathering tools resulted in 62 or fewer, that trait becomes extinct.

Step 7: Have students display data for the class. Facilitate class discussion.

- Which tool was easiest to use? Which tool collected the most beans? How did using a tool compare to using hands for bean-gathering?
- Which beans were the easiest to pick up? Some beans may be harder to spot because of their color. This information could be used to track the survivability of a type of bean.

	Spoon	Fork	Knife	Hand
Generation 1				
Generation 2				
Generation 3				
Generation 4				
Generation 5				

	Navy	Kidney	Lentil	Dried Pea
Generation 1				
Generation 2				
Generation 3				
Generation 4				
Generation 5				

This activity combines gathering and manipulating data and genetics. It's also easy to practice decimals, percents, and fractions using the data.

For more information and lessons on data analysis, probability, and statistics, see *Math Sense 3: Focus on Analysis*, at newreaderspress.com.

CCR Content Progressions[1]

This document contains the College and Career Readiness Standards for Adult Education, organized by level and also by domain. Standards associated with the Major Work of the Level (MWOTL) are identified in plain type and standards that support the MWOTL are in italics.

Number and Ratios: Understanding and Operations
Level A — Number Base Ten
1.NBT.2 Understand that the two digits of a two-digit number represent amounts of tens and ones. Understand the following as special cases: a. 10 can be thought of as a bundle of ten ones — called a "ten." b. The numbers from 11 to 19 are composed of a ten and one, two, three, four, five, six, seven, eight, or nine ones. c. The numbers 10, 20, 30, 40, 50, 60, 70, 80, 90 refer to one, two, three, four, five, six, seven, eight, or nine tens (and 0 ones).
1.NBT.3 Compare two two-digit numbers based on meanings of the tens and ones digits, recording the results of comparisons with the symbols >, =, and <.
1.NBT.4 Add within 100, including adding a two-digit number and a one-digit number, and adding a two-digit number and a multiple of 10, using concrete models or drawings and strategies based on place value, properties of operations, and/or the relationship between addition and subtraction; relate the strategy to a written method and explain the reasoning used. Understand that in adding two-digit numbers, one adds tens and tens, ones and ones; and sometimes it is necessary to compose a ten.
1.NBT.5 Given a two-digit number, mentally find 10 more or 10 less than the number, without having to count; explain the reasoning used.
1.NBT.6 Subtract multiples of 10 in the range 10-90 from multiples of 10 in the range 10-90 (positive or zero differences), using concrete models or drawings and strategies based on place value, properties of operations, and/or the relationship between addition and subtraction; relate the strategy to a written method and explain the reasoning used.

---

1   Critical content as defined by the Publishers' Criteria for the Common Core State Standards for Mathematics. Washington, DC: Author. Accessed January 18, 2015: http://www.corestandards.org/wp-content/uploads/Math_Publishers_ Criteria_K-8_Spring_2013_FINAL1.pdf and http://www.corestandards.org/wp-content/uploads/Math_Publishers_Criteria_ HS_Spring_2013_FINAL1.pdf

Level B — Number Base Ten

2.NBT.1 Understand that the three digits of a three-digit number represent amounts of hundreds, tens, and ones; e.g., 706 equals 7 hundreds, 0 tens, and 6 ones. Understand the following as special cases:

a. 100 can be thought of as a bundle of ten tens — called a "hundred."

b. The numbers 100, 200, 300, 400, 500, 600, 700, 800, 900 refer to one, two, three, four, five, six, seven, eight, or nine hundreds (and 0 tens and 0 ones).

2.NBT.2 Count within 1000; skip-count by 5s, 10s, and 100s.

2.NBT.3 Read and write numbers to 1000 using base-ten numerals, number names, and expanded form.

2.NBT.4 Compare two three-digit numbers based on meanings of the hundreds, tens, and ones digits, using >, =, and < symbols to record the results of comparisons.

2.NBT.6 *Add up to four two-digit numbers using strategies based on place value and properties of operations.*

2.NBT.7 Add and subtract within 1000, using concrete models or drawings and strategies based on place value, properties of operations, and/or the relationship between addition and subtraction; relate the strategy to a written method. Understand that in adding or subtracting three-digit numbers, one adds or subtracts hundreds and hundreds, tens and tens, ones and ones; and sometimes it is necessary to compose or decompose tens or hundreds.

2.NBT.8 Mentally add 10 or 100 to a given number 100–900, and mentally subtract 10 or 100 from a given number 100–900.

2.NBT.9 Explain why addition and subtraction strategies work, using place value and the properties of operations.

3.NBT.1 *Use place value understanding to round whole numbers to the nearest 10 or 100.*

3.NBT.2 Fluently add and subtract within 1000 using strategies and algorithms based on place value, properties of operations, and/or the relationship between addition and subtraction.

3.NBT.3 Multiply one-digit whole numbers by multiples of 10 in the range 10–90 (e.g., 9 × 80, 5 × 60) using strategies based on place value and properties of operations.

Level B — Fractions

3.NF.1 Understand a fraction 1/*b* as the quantity formed by 1 part when a whole is partitioned into b equal parts; understand a fraction *a*/*b* as the quantity formed by a parts of size 1/*b*.

3.NF.2 Understand a fraction as a number on the number line; represent fractions on a number line diagram.

3.NF.2a Represent a fraction 1/*b* on a number line diagram by defining the interval from 0 to 1 as the whole and partitioning it into b equal parts. Recognize that each part has size 1/*b* and that the endpoint of the part based at 0 locates the number 1/*b* on the number line.

3.NF.2b Represent a fraction *a*/*b* on a number line diagram by marking off *a* lengths 1/*b* from 0. Recognize that the resulting interval has size *a*/*b* and that its endpoint locates the number *a*/*b* on the number line.

3.NF.3 *Explain equivalence of fractions in special cases, and compare fractions by reasoning about their size.*

3.NF.3a *Understand two fractions as equivalent (equal) if they are the same size, or the same point on a number line.*

3.NF.3b *Recognize and generate simple equivalent fractions, e.g., 1/2 = 2/4, 4/6 = 2/3. Explain why the fractions are equivalent, e.g., by using a visual fraction model.*

3.NF.3c *Express whole numbers as fractions, and recognize fractions that are equivalent to whole numbers. Examples: Express 3 in the form 3 = 3/1; recognize that 6/1 = 6; locate 4/4 and 1 at the same point of a number line diagram.*

3.NF.3d *Compare two fractions with the same numerator or the same denominator by reasoning about their size. Recognize that comparisons are valid only when the two fractions refer to the same whole. Record the results of comparisons with the symbols >, =, or <, and justify the conclusions, e.g., by using a visual fraction model.*

Level C — Number Base Ten

4.NBT.1 *Recognize that in a multi-digit whole number, a digit in one place represents ten times what it represents in the place to its right. For example, recognize that 700 ÷ 70 = 10 by applying concepts of place value and division.*

4.NBT.2 *Read and write multi-digit whole numbers using base-ten numerals, number names, and expanded form. Compare two multi-digit numbers based on meanings of the digits in each place, using >, =, and < symbols to record the results of comparisons.*

4.NBT.3 *Use place value understanding to round multi-digit whole numbers to any place.*

4.NBT.4 Fluently add and subtract multi-digit whole numbers using the standard algorithm.

4.NBT.5 Multiply a whole number of up to four digits by a one-digit whole number, and multiply two two-digit numbers, using strategies based on place value and the properties of operations. Illustrate and explain the calculation by using equations, rectangular arrays, and/or area models.

4.NBT.6 Find whole-number quotients and remainders with up to four-digit dividends and one-digit divisors, using strategies based on place value, the properties of operations, and/or the relationship between multiplication and division. Illustrate and explain the calculation by using equations, rectangular arrays, and/or area models.

5.NBT.1 *Recognize that in a multi-digit number, a digit in one place represents 10 times as much as it represents in the place to its right and 1/10 of what it represents in the place to its left.*

5.NBT.2 *Explain patterns in the number of zeros of the product when multiplying a number by powers of 10, and explain patterns in the placement of the decimal point when a decimal is multiplied or divided by a power of 10. Use whole-number exponents to denote powers of 10.*

5.NBT.3 Read, write, and compare decimals to thousandths.

5.NBT.3a Read and write decimals to thousandths using base-ten numerals, number names, and expanded form, e.g.,
$347.392 = 3 \times 100 + 4 \times 10 + 7 \times 1 + 3 \times (1/10) + 9 \times (1/100) + 2 \times (1/1000)$.

5.NBT.3b Compare two decimals to thousandths based on meanings of the digits in each place, using >, =, and < symbols to record the results of comparisons.

5.NBT.4 *Use place value understanding to round decimals to any place.*

5.NBT.5 Fluently multiply multi-digit whole numbers using the standard algorithm. [NOTE: A "standard algorithm" might be any accepted algorithm that fits the experience and needs of the students.]

5.NBT.6 Find whole-number quotients of whole numbers with up to four-digit dividends and two-digit divisors, using strategies based on place value, the properties of operations, and/or the relationship between multiplication and division. Illustrate and explain the calculation by using equations, rectangular arrays, and/or area models.

5.NBT.7 Add, subtract, multiply, and divide decimals to hundredths, using concrete models or drawings and strategies based on place value, properties of operations, and/or the relationship between addition and subtraction; relate the strategy to a written method and explain the reasoning used. [NOTE: Applications involving financial literacy should be used.]

6.NS.2 Fluently divide multi-digit numbers using the standard algorithm.

6.NS.3 Fluently add, subtract, multiply, and divide multi-digit decimals using the standard algorithm for each operation.

6.NS.4 *Find the greatest common factor of two whole numbers less than or equal to 100 and the least common multiple of two whole numbers less than or equal to 12. Use the distributive property to express a sum of two whole numbers 1–100 with a common factor as a multiple of a sum of two whole numbers with no common factor. For example, express 36 + 8 as 4 (9 + 2).*

Level C — Fractions

4.NF.1 Explain why a fraction a/b is equivalent to a fraction $(n \times a)/(n \times b)$ by using visual fraction models, with attention to how the number and size of the parts differ even though the two fractions themselves are the same size. Use this principle to recognize and generate equivalent fractions.

4.NF.2 Compare two fractions with different numerators and different denominators, e.g., by creating common denominators or numerators, or by comparing to a benchmark fraction such as 1/2. Recognize that comparisons are valid only when the two fractions refer to the same whole. Record the results of comparisons with symbols >, =, or <, and justify the conclusions, e.g., by using a visual fraction model.

4.NF.3 *Understand a fraction a/b with a > 1 as a sum of fractions 1/b.*

4.NF.3a *Understand addition and subtraction of fractions as joining and separating parts referring to the same whole.*

4.NF.3b *Decompose a fraction into a sum of fractions with the same denominator in more than one way, recording each decomposition by an equation. Justify decompositions, e.g., by using a visual fraction model. Examples: 3/8 = 1/8 + 1/8 + 1/8; 3/8 = 1/8 + 2/8; 2 1/8 = 1 + 1 + 1/8 = 8/8 + 8/8 + 1/8*

4.NF.3c Add and subtract mixed numbers with like denominators, e.g., by replacing each mixed number with an equivalent fraction, and/or by using properties of operations and the relationship between addition and subtraction.

4.NF.3d Solve word problems involving addition and subtraction of fractions referring to the same whole and having like denominators, e.g., by using visual fraction models and equations to represent the problem.

4.NF.4 Apply and extend previous understandings of multiplication to multiply a fraction by a whole number.

4.NF.4a Understand a fraction a/b as a multiple of 1/b. For example, use a visual fraction model to represent 5/4 as the product 5 × (1/4), recording the conclusion by the equation 5/4 = 5 × (1/4).

4.NF.4b Understand a multiple of *a/b* as a multiple of 1/*b*, and use this understanding to multiply a fraction by a whole number. For example, use a visual fraction model to express 3 × (2/5) as 6 × (1/5), recognizing this product as 6/5. (In general, *n* × (*a/b*) = (*n* × *a*)/*b*.)

4.NF.4c Solve word problems involving multiplication of a fraction by a whole number, e.g., by using visual fraction models and equations to represent the problem. For example, if each person at a party will eat 3/8 of a pound of roast beef, and there will be 5 people at the party, how many pounds of roast beef will be needed? Between what two whole numbers does your answer lie?

4.NF.6 *Use decimal notation for fractions with denominators 10 or 100. For example, rewrite 0.62 as 62/100; describe a length as 0.62 meters; locate 0.62 on a number line diagram.*

4.NF.7 *Compare two decimals to hundredths by reasoning about their size. Recognize that comparisons are valid only when the two decimals refer to the same whole. Record the results of comparisons with the symbols >, =, or <, and justify the conclusions, e.g., by using a visual model.*

5.NF.1 Add and subtract fractions with unlike denominators (including mixed numbers) by replacing given fractions with equivalent fractions in such a way as to produce an equivalent sum or difference of fractions with like denominators. For example, 2/3 + 5/4 = 8/12 + 15/12 = 23/12. (In general, *a/b* + *c/d* = (*ad* + *bc*)/*bd*.)

5.NF.2 *Solve word problems involving addition and subtraction of fractions referring to the same whole, including cases of unlike denominators, e.g., by using visual fraction models or equations to represent the problem. Use benchmark fractions and number sense of fractions to estimate mentally and assess the reasonableness of answers. For example, recognize an incorrect result 2/5 + 1/2 = 3/7, by observing that 3/7 < 1/2.*

5.NF.3 Interpret a fraction as division of the numerator by the denominator (*a/b* = *a* ÷ *b*). Solve word problems involving division of whole numbers leading to answers in the form of fractions or mixed numbers, e.g., by using visual fraction models or equations to represent the problem. For example, interpret 3/4 as the result of dividing 3 by 4, noting that 3/4 multiplied by 4 equals 3, and that when 3 wholes are shared equally among 4 people each person has a share of size 3/4. If 9 people want to share a 50-pound sack of rice equally by weight, how many pounds of rice should each person get? Between what two whole numbers does your answer lie?

5.NF.4 Apply and extend previous understandings of multiplication to multiply a fraction or whole number by a fraction.

5.NF.5 *Interpret multiplication as scaling (resizing), by:*
a. *Comparing the size of a product to the size of one factor on the basis of the size of the other factor, without performing the indicated multiplication.*
b. *Explaining why multiplying a given number by a fraction greater than 1 results in a product greater than the given number (recognizing multiplication by whole numbers greater than 1 as a familiar case); explaining why multiplying a given number by a fraction less than 1 results in a product smaller than the given number; and relating the principle of fraction equivalence a/b = (nxa)/(nxb) to the effect of multiplying a/b by 1.*

5.NF.6 *Solve real world problems involving multiplication of fractions and mixed numbers, e.g., by using visual fraction models or equations to represent the problem.*

5.NF.7 Apply and extend previous understandings of division to divide unit fractions by whole numbers and whole numbers by unit fractions.

5.NF.7a Interpret division of a unit fraction by a non-zero whole number, and compute such quotients. For example, create a story context for (1/3) ÷ 4, and use a visual fraction model to show the quotient. Use the relationship between multiplication and division to explain that (1/3) ÷ 4 = 1/12 because (1/12) × 4 = 1/3.

5.NF.7b Interpret division of a whole number by a unit fraction, and compute such quotients. For example, create a story context for 4 ÷ (1/5), and use a visual fraction model to show the quotient. Use the relationship between multiplication and division to explain that 4 ÷ (1/5) = 20 because 20 × (1/5) = 4.

5.NF.7c *Solve real world problems involving division of unit fractions by non-zero whole numbers and division of whole numbers by unit fractions, e.g., by using visual fraction models and equations to represent the problem. For example, how much chocolate will each person get if 3 people share 1/2 lb of chocolate equally? How many 1/3-cup servings are in 2 cups of raisins?*

6.NS.1 *Interpret and compute quotients of fractions, and solve word problems involving division of fractions by fractions, e.g., by using visual fraction models and equations to represent the problem. For example, create a story context for (2/3) ÷ (3/4) and use a visual fraction model to show the quotient; use the relationship between multiplication and division to explain that (2/3) ÷ (3/4) = 8/9 because 3/4 of 8/9 is 2/3. (In general, (a/b) ÷ (c/d) = ad/bc.) How much chocolate will each person get if 3 people share 1/2 lb of chocolate equally? How many 3/4-cup servings are in 2/3 of a cup of yogurt? How wide is a rectangular strip of land with length 3/4 mi and area 1/2 square mi?*

6.RP.1 Understand the concept of a ratio and use ratio language to describe a ratio relationship between two quantities. For example, "The ratio of wings to beaks in the bird house at the zoo was 2:1, because for every 2 wings there was 1 beak." "For every vote candidate A received, candidate C received nearly three votes."

6.RP.2 Understand the concept of a unit rate a/b associated with a ratio *a:b* with *b* ≠ 0, and use rate language in the context of a ratio relationship. For example, "This recipe has a ratio of 3 cups of flour to 4 cups of sugar, so there is 3/4 cup of flour for each cup of sugar." "We paid $75 for 15 hamburgers, which is a rate of $5 per hamburger."

Level D — Number Systems

6.NS.5 *Understand that positive and negative numbers are used together to describe quantities having opposite directions or values (e.g., temperature above/below zero, elevation above/below sea level, credits/debits, positive/negative electric charge); use positive and negative numbers to represent quantities in real-world contexts, explaining the meaning of 0 in each situation.*

6.NS.6 *Understand a rational number as a point on the number line. Extend number line diagrams and coordinate axes familiar from previous grades to represent points on the line and in the plane with negative number coordinates.*

6.NS.6a *Recognize opposite signs of numbers as indicating locations on opposite sides of 0 on the number line; recognize that the opposite of the opposite of a number is the number itself, e.g., −(−3) = 3, and that 0 is its own opposite.*

6.NS.6b *Understand signs of numbers in ordered pairs as indicating locations in quadrants of the coordinate plane; recognize that when two ordered pairs differ only by signs, the locations of the points are related by reflections across one or both axes.*

6.NS.6c *Find and position integers and other rational numbers on a horizontal or vertical number line diagram; find and position pairs of integers and other rational numbers on a coordinate plane.*

6.NS.7 Understand a rational number as a point on the number line. Extend number line diagrams and coordinate axes familiar from previous grades to represent points on the line and in the plane with negative number coordinates.

6.NS.7a *Interpret statements of inequality as statements about the relative position of two numbers on a number line diagram. For example, interpret –3 > –7 as a statement that –3 is located to the right of –7 on a number line oriented from left to right.*

6.NS.7b *Write, interpret, and explain statements of order for rational numbers in real-world contexts. For example, write –3° C > –7° C to express the fact that –3 degrees C is warmer than –7 degrees C.*

6.NS.7c *Understand the absolute value of a rational number as its distance from 0 on the number line; interpret absolute value as magnitude for a positive or negative quantity in a real-world situation. For example, for an account balance of –30 dollars, write |–30| = 30 to describe the size of the debt in dollars.*

6.NS.7d *Distinguish comparisons of absolute value from statements about order. For example, recognize that an account balance less than –30 dollars represents a debt greater than 30 dollars.*

6.NS.8 *Solve real-world and mathematical problems by graphing points in all four quadrants of the coordinate plane. Include use of coordinates and absolute value to find distances between points with the same first coordinate or the same second coordinate.*

7.NS.1 Apply and extend previous understandings of addition and subtraction to add and subtract rational numbers; represent addition and subtraction on a horizontal or vertical number line diagram.

7.NS.1a *Describe situations in which opposite quantities combine to make 0. For example, a hydrogen atom has 0 charge because its two constituents are oppositely charged.*

7.NS.1b Understand $p + q$ as the number located a distance $|q|$ from p, in the positive or negative direction depending on whether q is positive or negative. Show that a number and its opposite have a sum of 0 (are additive inverses). Interpret sums of rational numbers by describing real-world contexts.

7.NS.1c Understand subtraction of rational numbers as adding the additive inverse, $p - q = p + (-q)$. Show that the distance between two rational numbers on the number line is the absolute value of their difference, and apply this principle in real-world contexts.

7.NS.1d Apply properties of operations as strategies to add and subtract rational numbers.

7.NS.2 Apply and extend previous understandings of multiplication and division and of fractions to multiply and divide rational numbers.

7.NS.2a Understand that multiplication is extended from fractions to rational numbers by requiring that operations continue to satisfy the properties of operations, particularly the distributive property, leading to products such as $(-1)(-1) = 1$ and the rules for multiplying signed numbers. Interpret products of rational numbers by describing real-world contexts.

7.NS.2b Understand that integers can be divided, provided that the divisor is not zero, and every quotient of integers (with non-zero divisor) is a rational number. If p and q are integers, then $-(p/q) = (-p)/q = p/(-q)$. Interpret quotients of rational numbers by describing real-world contexts.

7.NS.2c Apply properties of operations as strategies to multiply and divide rational numbers.
7.NS.2d Convert a rational number to a decimal using long division; know that the decimal form of a rational number terminates in 0s or eventually repeats.
7.NS.3 Solve real-world and mathematical problems involving the four operations with rational numbers.
8.NS.2 *Use rational approximations of irrational numbers to compare the size of irrational numbers, locate them approximately on a number line diagram, and estimate the value of expressions (e.g., π^2). For example, by truncating the decimal expansion of $\sqrt{2}$, show that $\sqrt{2}$ is between 1 and 2, then between 1.4 and 1.5, and explain how to continue on to get better approximations.*

Level D — Ratio and Proportional Reasoning

6.RP.3 Use ratio and rate reasoning to solve real-world and mathematical problems, e.g., by reasoning about tables of equivalent ratios, tape diagrams, double number line diagrams, or equations.
6.RP.3a Make tables of equivalent ratios relating quantities with whole-number measurements, find missing values in the tables, and plot the pairs of values on the coordinate plane. Use tables to compare ratios.
6.RP.3b Solve unit rate problems including those involving unit pricing and constant speed. For example, if it took 7 hours to mow 4 lawns, then at that rate, how many lawns could be mowed in 35 hours? At what rate were lawns being mowed?
6.RP.3c Find a percent of a quantity as a rate per 100 (e.g., 30% of a quantity means 30/100 times the quantity); solve problems involving finding the whole, given a part and the percent.
6.RP.3d Use ratio reasoning to convert measurement units; manipulate and transform units appropriately when multiplying or dividing quantities.
7.RP.1 Compute unit rates associated with ratios of fractions, including ratios of lengths, areas and other quantities measured in like or different units. For example, if a person walks 1/2 mile in each 1/4 hour, compute the unit rate as the complex fraction (1/2)/(1/4) miles per hour, equivalently 2 miles per hour.
7.RP.2 Recognize and represent proportional relationships between quantities.
7.RP.2a Decide whether two quantities are in a proportional relationship, e.g., by testing for equivalent ratios in a table or graphing on a coordinate plane and observing whether the graph is a straight line through the origin.
7.RP.2b Identify the constant of proportionality (unit rate) in tables, graphs, equations, diagrams, and verbal descriptions of proportional relationships. (Also see 8.EE.5)
7.RP.2c Represent proportional relationships by equations. For example, if total cost t is proportional to the number n of items purchased at a constant price p, the relationship between the total cost and the number of items can be expressed as $t = pn$.
7.RP.2d Explain what a point (x, y) on the graph of a proportional relationship means in terms of the situation, with special attention to the points $(0, 0)$ and $(1, r)$ where r is the unit rate.
7.RP.3 Use proportional relationships to solve multistep ratio and percent problems. Examples: simple interest, tax, markups and markdowns, gratuities and commissions, fees, percent increase and decrease, percent error. (Also see 7.G.1 and G.MG.2)

Level E — Number and Quantity

N.RN.2 Rewrite expressions involving radicals and rational exponents using the properties of exponents.

N.Q.1 Use units as a way to understand problems and to guide the solution of multi-step problems; choose and interpret units consistently in formulas; choose and interpret the scale and the origin in graphs and data displays.

N.Q.3 Choose a level of accuracy appropriate to limitations on measurement when reporting quantities.

Algebra and Functions

Level A — Operations and Algebraic Thinking

1.OA.2 *Solve word problems that call for addition of three whole numbers whose sum is less than or equal to 20, e.g., by using objects, drawings, and equations with a symbol for the unknown number to represent the problem.*

1.OA.3 Apply properties of operations as strategies to add and subtract. Examples: If $8 + 3 = 11$ is known, then $3 + 8 = 11$ is also known. (Commutative property of addition.) To add $2 + 6 + 4$, the second two numbers can be added to make a ten, so $2 + 6 + 4 = 2 + 10 = 12$. (Associative property of addition.)

1.OA.4 Understand subtraction as an unknown-addend problem. For example, subtract $10 - 8$ by finding the number that makes 10 when added to 8.

1.OA.5 Relate counting to addition and subtraction (e.g., by counting on 2 to add 2).

1.OA.6 Add and subtract within 20, demonstrating fluency for addition and subtraction within 10. Use strategies such as counting on; making ten (e.g., $8 + 6 = 8 + 2 + 4 = 10 + 4 = 14$); decomposing a number leading to a ten (e.g., $13 - 4 = 13 - 3 - 1 = 10 - 1 = 9$); using the relationship between addition and subtraction (e.g., knowing that $8 + 4 = 12$, one knows $12 - 8 = 4$); and creating equivalent but easier or known sums (e.g., adding $6 + 7$ by creating the known equivalent $6 + 6 + 1 = 12 + 1 = 13$).

1.OA.7 *Understand the meaning of the equal sign, and determine if equations involving addition and subtraction are true or false. For example, which of the following equations are true and which are false? $6 = 6$, $7 = 8 - 1$, $5 + 2 = 2 + 5$, $4 + 1 = 5 + 2$.*

1.OA.8 Determine the unknown whole number in an addition or subtraction equation relating three whole numbers. For example, determine the unknown number that makes the equation true in each of the equations $8 + ? = 11$, $5 = \_ - 3$, $6 + 6 = \_$.

Level B — Operations and Algebraic Thinking

2.OA.1 Use addition and subtraction within 100 to solve one- and two-step word problems involving situations of adding to, taking from, putting together, taking apart, and comparing, with unknowns in all positions, e.g., by using drawings and equations with a symbol for the unknown number to represent the problem.

2.OA.2 Fluently add and subtract within 20 using mental strategies. By end of Grade 2, know from memory all sums of two one-digit numbers.

3.OA.1 Interpret products of whole numbers, e.g., interpret 5×7 as the total number of objects in 5 groups of 7 objects each. For example, describe a context in which a total number of objects can be expressed as 5×7.

3.OA.2 Interpret whole-number quotients of whole numbers, e.g., interpret 56×8 as the number of objects in each share when 56 objects are partitioned equally into 8 shares, or as a number of shares when 56 objects are partitioned into equal shares of 8 objects each. For example, describe a context in which a number of shares or a number of groups can be expressed as $56 \div 8$.

3.OA.3 Use multiplication and division within 100 to solve word problems in situations involving equal groups, arrays, and measurement quantities, e.g., by using drawings and equations with a symbol for the unknown number to represent the problem.

3.OA.4 Determine the unknown whole number in a multiplication or division equation relating three whole numbers. For example, determine the unknown number that makes the equation true in each of the equations $8 \times ? = 48$, $5 = \_ \div 3$, $6 \times 6 = ?$

3.OA.5 *Apply properties of operations as strategies to multiply and divide. Examples: If $6 \times 4 = 24$ is known, then $4 \times 6 = 24$ is also known. (Commutative property of multiplication.) $3 \times 5 \times 2$ can be found by $3 \times 5 = 15$, then $15 \times 2 = 30$, or by $5 \times 2 = 10$, then $3 \times 10 = 30$. (Associative property of multiplication.) Knowing that $8 \times 5 = 40$ and $8 \times 2 = 16$, one can find 8×7 as $8 \times (5 + 2) = (8 \times 5) + (8 \times 2) = 40 + 16 = 56$. (Distributive property.)*

3.OA.6 Understand division as an unknown-factor problem. For example, find $32 \div 8$ by finding the number that makes 32 when multiplied by 8.

3.OA.7 *Fluently multiply and divide within 100, using strategies such as the relationship between multiplication and division (e.g., knowing that $8 \times 5 = 40$, one knows $40 \div 5 = 8$) or properties of operations. By the end of Grade 3, know from memory all products of two one-digit numbers.*

3.OA.8 *Solve two-step word problems using the four operations. Represent these problems using equations with a letter standing for the unknown quantity. Assess the reasonableness of answers using mental computation and estimation strategies including rounding.*

3.OA.9 *Identify arithmetic patterns (including patterns in the addition table or multiplication table), and explain them using properties of operations. For example, observe that 4 times a number is always even, and explain why 4 times a number can be decomposed into two equal addends.*

Level C — Operations and Algebraic Thinking

4.OA.1 Interpret a multiplication equation as a comparison, e.g., interpret $35 = 5 \times 7$ as a statement that 35 is 5 times as many as 7 and 7 times as many as 5. Represent verbal statements of multiplicative comparisons as multiplication equations.

4.OA.2 *Multiply or divide to solve word problems involving multiplicative comparison, e.g., by using drawings and equations with a symbol for the unknown number to represent the problem, distinguishing multiplicative comparison from additive comparison.*

4.OA.3 *Solve multistep word problems posed with whole numbers and having whole-number answers using the four operations, including problems in which remainders must be interpreted. Represent these problems using equations with a letter standing for the unknown quantity. Assess the reasonableness of answers using mental computation and estimation strategies including rounding.*

4.OA.4 *Find all factor pairs for a whole number in the range 1–100. Recognize that a whole number is a multiple of each of its factors. Determine whether a given whole number in the range 1–100 is a multiple of a given one-digit number. Determine whether a given whole number in the range 1–100 is prime or composite.*

4.OA.5 Generate a number or shape pattern that follows a given rule. Identify apparent features of the pattern that were not explicit in the rule itself. For example, given the rule "Add 3" and the starting number 1, generate terms in the resulting sequence and observe that the terms appear to alternate between odd and even numbers. Explain informally why the numbers will continue to alternate in this way.

5.OA.1 Use parentheses, brackets, or braces in numerical expressions, and evaluate expressions with these symbols.

5.OA.2 Write simple expressions that record calculations with numbers, and interpret numerical expressions without evaluating them. For example, express the calculation "add 8 and 7, then multiply by 2" as $2 \times (8 + 7)$. Recognize that $3 \times (18932 + 921)$ is three times as large as $18932 + 921$, without having to calculate the indicated sum or product.

6.EE.1 Write and evaluate numerical expressions involving whole-number exponents.

6.EE.2 Write, read, and evaluate expressions in which letters stand for numbers.

6.EE.2a Write expressions that record operations with numbers and with letters standing for numbers. For example, express the calculation "Subtract y from 5" as $5 - y$.

6.EE.2b Identify parts of an expression using mathematical terms (sum, term, product, factor, quotient, coefficient); view one or more parts of an expression as a single entity. For example, describe the expression $2 (8 + 7)$ as a product of two factors; view $(8 + 7)$ as both a single entity and a sum of two terms.

6.EE.2c Evaluate expressions at specific values of their variables. Include expressions that arise from formulas used in real-world problems. Perform arithmetic operations, including those involving whole-number exponents, in the conventional order when there are no parentheses to specify a particular order (Order of Operations). For example, use the formulas $V = s^3$ and $A = 6 s^2$ to find the volume and surface area of a cube with sides of length $s = 1/2$.

6.EE.3 *Apply the properties of operations to generate equivalent expressions. For example, apply the distributive property to the expression $3(2 + x)$ to produce the equivalent expression $6 + 3x$; apply the distributive property to the expression $24x + 18y$ to produce the equivalent expression $6(4x + 3y)$; apply properties of operations to $y + y + y$ to produce the equivalent expression $3y$.*

6.EE.4 *Identify when two expressions are equivalent (i.e., when the two expressions name the same number regardless of which value is substituted into them). For example, the expressions $y + y + y$ and $3y$ are equivalent because they name the same number regardless of which number y stands for.*

6.EE.5 *Understand solving an equation or inequality as a process of answering a question: which values from a specified set, if any, make the equation or inequality true? Use substitution to determine whether a given number in a specified set makes an equation or inequality true.*

6.EE.6 Use variables to represent numbers and write expressions when solving a real-world or mathematical problem; understand that a variable can represent an unknown number, or, depending on the purpose at hand, any number in a specified set.

6.EE.7 *Solve real-world and mathematical problems by writing and solving equations of the form $x + p = q$ and $px = q$ for cases in which p, q and x are all nonnegative rational numbers.*

6.EE.8 *Write an inequality of the form $x > c$ or $x < c$ to represent a constraint or condition in a real-world or mathematical problem. Recognize that inequalities of the form $x > c$ or $x < c$ have infinitely many solutions; represent solutions of such inequalities on number line diagrams.*

6.EE.9 Use variables to represent two quantities in a real-world problem that change in relationship to one another; write an equation to express one quantity, thought of as the dependent variable, in terms of the other quantity, thought of as the independent variable. Analyze the relationship between the dependent and independent variables using graphs and tables, and relate these to the equation. *For example, in a problem involving motion at constant speed, list and graph ordered pairs of distances and times, and write the equation $d = 65t$ to represent the relationship between distance and time.*

Level D — Expressions and Equations

7.EE.1 Apply properties of operations as strategies to add, subtract, factor, and expand linear expressions with rational coefficients.

7.EE.2 *Understand that rewriting an expression in different forms in a problem context can shed light on the problem and how the quantities in it are related. For example, $a + 0.05a = 1.05a$ means that "increase by 5%" is the same as "multiply by 1.05." [Also see A.SSE.2, A.SSE .3, A.SSE .3a, A.CED.4]*

7.EE.3 Solve multi-step real-life and mathematical problems posed with positive and negative rational numbers in any form (whole numbers, fractions, and decimals), using tools strategically. Apply properties of operations to calculate with numbers in any form; convert between forms as appropriate; and assess the reasonableness of answers using mental computation and estimation strategies. For example: If a woman making $25 an hour gets a 10% raise, she will make an additional 1/10 of her salary an hour, or $2.50, for a new salary of $27.50. If you want to place a towel bar 9 3/4 inches long in the center of a door that is 27 1/2 inches wide, you will need to place the bar about 9 inches from each edge; this estimate can be used as a check on the exact computation.

7.EE.4 Use variables to represent quantities in a real-world or mathematical problem, and construct simple equations and inequalities to solve problems by reasoning about the quantities.

7.EE.4a Solve word problems leading to equations of the form $px + q = r$ and $p(x + q) = r$, where p, q, and r are specific rational numbers. Solve equations of these forms fluently. Compare an algebraic solution to an arithmetic solution, identifying the sequence of the operations used in each approach. For example, the perimeter of a rectangle is 54 cm. Its length is 6 cm. What is its width?

7.EE.4b Solve word problems leading to inequalities of the form $px + q > r$ or $px + q < r$, where p, q, and r are specific rational numbers. Graph the solution set of the inequality and interpret it in the context of the problem. *For example: As a salesperson, you are paid $50 per week plus $3 per sale. This week you want your pay to be at least $100. Write an inequality for the number of sales you need to make, and describe the solutions.*

8.EE.1. *Know and apply the properties of integer exponents to generate equivalent numerical expressions. For example, $3^2 \times 3^{-5} = 3^{-3} = (1/3)^3 = 1/27$.*

8.EE.2 *Use square root and cube root symbols to represent solutions to equations of the form $x^2 = p$ and $x^3 = p$, where p is a positive rational number. Evaluate square roots of small perfect squares and cube roots of small perfect cubes. Know that $\sqrt{2}$ is irrational.*

8.EE.3 *Use numbers expressed in the form of a single digit times an integer power of 10 to estimate very large or very small quantities, and to express how many times as much one is than the other. For example, estimate the population of the United States as 3×10^8 and the population of the world as 7×10^9, and determine that the world population is more than 20 times larger.*

8.EE.4 *Perform operations with numbers expressed in scientific notation, including problems where both decimal and scientific notation are used. Use scientific notation and choose units of appropriate size for measurements of very large or very small quantities (e.g., use millimeters per year for seafloor spreading). Interpret scientific notation that has been generated by technology.*

8.EE.5 Graph proportional relationships, interpreting the unit rate as the slope of the graph. Compare two different proportional relationships represented in different ways. For example, compare a distance-time graph to a distance-time equation to determine which of two moving objects has greater speed. [Also see 7.RP.2b]

8.EE.7 Solve linear equations in one variable.

8.EE.7a Give examples of linear equations in one variable with one solution, infinitely many solutions, or no solutions. Show which of these possibilities is the case by successively transforming the given equation into simpler forms, until an equivalent equation of the form $x = a$, $a = a$, or $a = b$ results (where a and b are different numbers).

8.EE.7b Solve linear equations with rational number coefficients, including equations whose solutions require expanding expressions using the distributive property and collecting like terms.

8.EE.8 Analyze and solve pairs of simultaneous linear equations.

8.EE.8a Understand that solutions to a system of two linear equations in two variables correspond to points of intersection of their graphs, because points of intersection satisfy both equations simultaneously.

8.EE.8b Solve systems of two linear equations in two variables algebraically, and estimate solutions by graphing the equations. Solve simple cases by inspection. For example, $3x + 2y = 5$ and $3x + 2y = 6$ have no solution because $3x + 2y$ cannot simultaneously be 5 and 6.

8.EE.8c Solve real-world and mathematical problems leading to two linear equations in two variables. For example, given coordinates for two pairs of points, determine whether the line through the first pair of points intersects the line through the second pair.

Level D — Functions

8.F.1 Understand that a function is a rule that assigns to each input exactly one output. The graph of a function is the set of ordered pairs consisting of an input and the corresponding output. [Also see F.IF.1]

8.F.3 Interpret the equation $y = mx + b$ as defining a linear function, whose graph is a straight line; give examples of functions that are not linear. For example, the function $A = s^2$ giving the area of a square as a function of its side length is not linear because its graph contains the points (1,1), (2,4) and (3,9), which are not on a straight line.

8.F.4 *Construct a function to model a linear relationship between two quantities. Determine the rate of change and initial value of the function from a description of a relationship or from two (x, y) values, including reading these from a table or from a graph. Interpret the rate of change and initial value of a linear function in terms of the situation it models, and in terms of its graph or a table of values. [Also see F.BF.1 and F.LE.5]*

8.F.5 *Describe qualitatively the functional relationship between two quantities by analyzing a graph (e.g., where the function is increasing or decreasing, linear or nonlinear). Sketch a graph that exhibits the qualitative features of a function that has been described verbally. [Also see A.REI.10 and F.IF.7]*

Level E — Algebra

A.SSE.1 Interpret expressions that represent a quantity in terms of its context.

A.SSE.1a Interpret parts of an expression, such as terms, factors, and coefficients.

A.SSE.2 Use the structure of an expression to identify ways to rewrite it. For example, see $x^4 - y^4$ as $(x^2)^2 - (y^2)^2$, thus recognizing it as a difference of squares that can be factored as $(x^2 - y^2)(x^2 + y^2)$. [Also see 7.EE.2]

A.SSE.3 Choose and produce an equivalent form of an expression to reveal and explain properties of the quantity represented by the expression. [Also see 7.EE.2]

A.SSE.3a Factor a quadratic expression to reveal the zeros of the function it defines. [Also see 7.EE.2]

A.APR.1 Understand that polynomials form a system analogous to the integers, namely, they are closed under the operations of addition, subtraction, and multiplication; add, subtract, and multiply polynomials. [NOTE: Emphasis should be on operations with polynomials.]

A.APR.6 Rewrite simple rational expressions in different forms; write $a(x)/b(x)$ in the form $q(x) + r(x)/b(x)$, where $a(x)$, $b(x)$, $q(x)$, and $r(x)$ are polynomials with the degree of $r(x)$ less than the degree of $b(x)$, using inspection, long division, or, for the more complicated examples, a computer algebra system.

A.CED.1 Create equations and inequalities in one variable and use them to solve problems. Include equations arising from linear and quadratic functions, and simple rational and exponential functions.

A.CED.2 Create equations in two or more variables to represent relationships between quantities; graph equations on coordinate axes with labels and scales.

A.CED.3 Represent constraints by equations or inequalities, and by systems of equations and/or inequalities, and interpret solutions as viable or non-viable options in a modeling context. For example, represent inequalities describing nutritional and cost constraints on combinations of different foods.

A.CED.4 *Rearrange formulas to highlight a quantity of interest, using the same reasoning as in solving equations. For example, rearrange Ohm's law V = IR to highlight resistance R.) [Also see 7.EE.2 and F.IF.8]*

A.REI.1 *Explain each step in solving a simple equation as following from the equality of numbers asserted at the previous step, starting from the assumption that the original equation has a solution. Construct a viable argument to justify a solution method.*

A.REI.2 *Solve simple rational and radical equations in one variable, and give examples showing how extraneous solutions may arise.*

A.REI.3 Solve linear equations and inequalities in one variable, including equations with coefficients represented by letters.

A.REI.4 Solve quadratic equations in one variable.
A.REI.6 *Solve systems of linear equations exactly and approximately (e.g., with graphs), focusing on pairs of linear equations in two variables.*
A.REI.10 *Understand that the graph of an equation in two variables is the set of all its solutions plotted in the coordinate plane, often forming a curve (which could be a line). [Also see 8.F.5]*

Level E — Functions

F.IF.1 *Understand that a function from one set (called the domain) to another set (called the range) assigns to each element of the domain exactly one element of the range. If f is a function and x is an element of its domain, then f(x) denotes the output of f corresponding to the input x. The graph of f is the graph of the equation y = f(x). [Also see 8.F.1]*
F.IF.2 *Use function notation, evaluate functions for inputs in their domains, and interpret statements that use function notation in terms of a context.*
F.IF.4 For a function that models a relationship between two quantities, interpret key features of graphs and tables in terms of the quantities, and sketch graphs showing key features given a verbal description of the relationship. Key features include: intercepts; intervals where the function is increasing, decreasing, positive, or negative; relative maximums and minimums; symmetries; end behavior; and periodicity. [NOTE: Key features to include: intercepts; intervals where the function is increasing, decreasing, positive, or negative; relative maximums and minimums; symmetries; end behavior; and periodicity.]
F.IF.5 Relate the domain of a function to its graph and, where applicable, to the quantitative relationship it describes. For example, if the function $h(n)$ gives the number of person-hours it takes to assemble n engines in a factory, then the positive integers would be an appropriate domain for the function.
F.IF.6 *Calculate and interpret the average rate of change of a function (presented symbolically or as a table) over a specified interval. Estimate the rate of change from a graph. [NOTE: See modeling conceptual categories]*
F.IF.7 Graph functions expressed symbolically and show key features of the graph, by hand in simple cases and using technology for more complicated cases. [Also see 8.F.5]
F.IF.8b Use the properties of exponents to interpret expressions for exponential functions. For example, identify percent rate of change in functions such as $y = (1.02)^t$, $y = (0.97)^t$, $y = (1.01)^{12t}$, $y = (1.2)^{(t/10)}$, and classify them as representing exponential growth or decay.
F.IF.9 Compare properties of two functions each represented in a different way (algebraically, graphically, numerically in tables, or by verbal descriptions). For example, given a graph of one quadratic function and an algebraic expression for another, say which has the larger maximum.
F.BF.1 Write a function that describes a relationship between two quantities. [Also see 8.F.4]
F.LE.1 Distinguish between situations that can be modeled with linear functions and with exponential functions.
F.LE.1b Recognize situations in which one quantity changes at a constant rate per unit interval relative to another.
F.LE.1c Recognize situations in which a quantity grows or decays by a constant percent rate per unit interval relative to another.

F.LE.5 Interpret the parameters in a linear or exponential function in terms of a context. [Also see 8.F.4]

Geometry

Level A — Geometry and Geometric Measurement

K.G.4 Analyze and compare two- and three-dimensional shapes, in different sizes and orientations, using informal language to describe their similarities, differences, parts (e.g., number of sides and vertices/"corners") and other attributes (e.g., having sides of equal length).

1.G.2 Compose two-dimensional shapes (rectangles, squares, trapezoids, triangles, half-circles, and quarter-circles) or three-dimensional shapes (cubes, right rectangular prisms, right circular cones, and right circular cylinders) to create a composite shape, and compose new shapes from the composite shape.

1.MD.2 Express the length of an object as a whole number of length units, by laying multiple copies of a shorter object (the length unit) end to end; understand that the length measurement of an object is the number of same-size length units that span it with no gaps or overlaps. Limit to contexts where the object being measured is spanned by a whole number of length units with no gaps or overlaps.

Level B — Geometry and Geometric Measurement

2.G.1 Recognize and draw shapes having specified attributes, such as a given number of angles or a given number of equal faces. Identify triangles, quadrilaterals, pentagons, hexagons, and cubes.

2.G.3 Partition circles and rectangles into two, three, or four equal shares, describe the shares using the words halves, thirds, half of, a third of, etc., and describe the whole as two halves, three thirds, four fourths. Recognize that equal shares of identical wholes need not have the same shape.

3.G.1 *Understand that shapes in different categories (e.g., rhombuses, rectangles, and others) may share attributes (e.g., having four sides), and that the shared attributes can define a larger category (e.g., quadrilaterals). Recognize rhombuses, rectangles, and squares as examples of quadrilaterals, and draw examples of quadrilaterals that do not belong to any of these subcategories.*

3.G.2 Partition shapes into parts with equal areas. Express the area of each part as a unit fraction of the whole. For example, partition a shape into 4 parts with equal area, and describe the area of each part as 1/4 of the area of the shape.

2.MD.2 *Measure the length of an object twice, using length units of different lengths for the two measurements; describe how the two measurements relate to the size of the unit chosen.*

2.MD.3 Estimate lengths using units of inches, feet, centimeters, and meters.

2.MD.4 Measure to determine how much longer one object is than another, expressing the length difference in terms of a standard length unit.

2.MD.6 *Represent whole numbers as lengths from 0 on a number line diagram with equally spaced points corresponding to the numbers 0, 1, 2, …, and represent whole-number sums and differences within 100 on a number line diagram.*

3.MD.1 Tell and write time to the nearest minute and measure time intervals in minutes. Solve word problems involving addition and subtraction of time intervals in minutes, e.g., by representing the problem on a number line diagram.

3.MD.2 Measure and estimate liquid volumes and masses of objects using standard units of grams (g), kilograms (kg), and liters (l). Add, subtract, multiply, or divide to solve one-step word problems involving masses or volumes that are given in the same units, e.g., by using drawings (such as a beaker with a measurement scale) to represent the problem.

3.MD.5 Recognize area as an attribute of plane figures and understand concepts of area measurement.

a. A square with side length 1 unit, called "a unit square," is said to have "one square unit" of area, and can be used to measure area.

b. A plane figure which can be covered without gaps or overlaps by n unit squares is said to have an area of *n* square units.

3.MD.6 Measure areas by counting unit squares (square cm, square m, square in, square ft, and improvised units).

3.MD.7 Relate area to the operations of multiplication and addition.

3.MD.7a Find the area of a rectangle with whole-number side lengths by tiling it, and show that the area is the same as would be found by multiplying the side lengths.

3.MD.7b Multiply side lengths to find areas of rectangles with whole-number side lengths in the context of solving real world and mathematical problems, and represent whole-number products as rectangular areas in mathematical reasoning.

3.MD.7c Use tiling to show in a concrete case that the area of a rectangle with whole-number side lengths a and $b + c$ is the sum of $a \times b$ and $a \times c$. Use area models to represent the distributive property in mathematical reasoning.

3.MD.7d Recognize area as additive. Find areas of rectilinear figures by decomposing them into non-overlapping rectangles and adding the areas of the non-overlapping parts, applying this technique to solve real world problems.

3.MD.8 *Solve real world and mathematical problems involving perimeters of polygons, including finding the perimeter given the side lengths, finding an unknown side length, and exhibiting rectangles with the same perimeter and different areas or with the same area and different perimeters.*

Level C — Geometry and Geometric Measurement

4.G.1 Draw points, lines, line segments, rays, angles (right, acute, obtuse), and perpendicular and parallel lines. Identify these in two-dimensional figures.

5.G.1 Use a pair of perpendicular number lines, called axes, to define a coordinate system, with the intersection of the lines (the origin) arranged to coincide with the 0 on each line and a given point in the plane located by using an ordered pair of numbers, called its coordinates. Understand that the first number indicates how far to travel from the origin in the direction of one axis, and the second number indicates how far to travel in the direction of the second axis, with the convention that the names of the two axes and the coordinates correspond (e.g., *x*-axis and *x*-coordinate, *y*-axis and *y*-coordinate).

5.G.2 Represent real world and mathematical problems by graphing points in the first quadrant of the coordinate plane, and interpret coordinate values of points in the context of the situation.

5.G.3 Understand that attributes belonging to a category of two-dimensional figures also belong to all subcategories of that category. For example, all rectangles have four right angles and squares are rectangles, so all squares have four right angles.

6.G.1 Find the area of right triangles, other triangles, special quadrilaterals, and polygons by composing into rectangles or decomposing into triangles and other shapes; apply these techniques in the context of solving real-world and mathematical problems.

6.G.3 Draw polygons in the coordinate plane given coordinates for the vertices; use coordinates to find the length of a side joining points with the same first coordinate or the same second coordinate. Apply these techniques in the context of solving real-world and mathematical problems.

6.G.4 *Represent three-dimensional figures using nets made up of rectangles and triangles, and use the nets to find the surface area of these figures. Apply these techniques in the context of solving real-world and mathematical problems.*

4.MD.2 *Use the four operations to solve word problems involving distances, intervals of time, liquid volumes, masses of objects, and money, including problems involving simple fractions or decimals, and problems that require expressing measurements given in a larger unit in terms of a smaller unit. Represent measurement quantities using diagrams such as number line diagrams that feature a measurement scale.*

4.MD.3 *Apply the area and perimeter formulas for rectangles in real world and mathematical problems. For example, find the width of a rectangular room given the area of the flooring and the length, by viewing the area formula as a multiplication equation with an unknown factor.*

4.MD.5 *Recognize angles as geometric shapes that are formed wherever two rays share a common endpoint, and understand concepts of angle measurement:*

a. *An angle is measured with reference to a circle with its center at the common endpoint of the rays, by considering the fraction of the circular arc between the points where the two rays intersect the circle. An angle that turns through 1/360 of a circle is called a "one-degree angle," and can be used to measure angles.*

b. *An angle that turns through n one-degree angles is said to have an angle measure of n degrees.*

4.MD.6 *Measure angles in whole-number degrees using a protractor. Sketch angles of specified measure.*

4.MD.7 *Recognize angle measure as additive. When an angle is decomposed into non-overlapping parts, the angle measure of the whole is the sum of the angle measures of the parts. Solve addition and subtraction problems to find unknown angles on a diagram in real world and mathematical problems, e.g., by using an equation with a symbol for the unknown angle measure.*

5.MD.1 *Convert among different-sized standard measurement units within a given measurement system (e.g., convert 5 cm to 0.05 m), and use these conversions in solving multi-step, real world problems.*

5.MD.3 Recognize volume as an attribute of solid figures and understand concepts of volume measurement.

a. A cube with side length 1 unit, called a "unit cube," is said to have "one cubic unit" of volume, and can be used to measure volume.

b. A solid figure which can be packed without gaps or overlaps using n unit cubes is said to have a volume of n cubic units.

5.MD.4 Measure volumes by counting unit cubes, using cubic cm, cubic in, cubic ft, and improvised units.

5.MD.5 Relate volume to the operations of multiplication and addition and solve real world and mathematical problems involving volume.

5.MD.5a Find the volume of a right rectangular prism with whole-number side lengths by packing it with unit cubes, and show that the volume is the same as would be found by multiplying the edge lengths, equivalently by multiplying the height by the area of the base. Represent threefold whole-number products as volumes, e.g., to represent the associative property of multiplication.

5.MD.5b Apply the formulas $V = l \times w \times h$ and $V = b \times h$ for rectangular prisms to find volumes of right rectangular prisms with whole-number edge lengths in the context of solving real world and mathematical problems.

5.MD.5c Recognize volume as additive. Find volumes of solid figures composed of two non-overlapping right rectangular prisms by adding the volumes of the non-overlapping parts, applying this technique to solve real world problems.

Level D — Geometry

7.G.1 Solve problems involving scale drawings of geometric figures, including computing actual lengths and areas from a scale drawing and reproducing a scale drawing at a different scale. [Also see 7.RP.3]

7.G.4 Know the formulas for the area and circumference of a circle and use them to solve problems; give an informal derivation of the relationship between the circumference and area of a circle.

7.G.5 *Use facts about supplementary, complementary, vertical, and adjacent angles in a multi-step problem to write and solve simple equations for an unknown angle in a figure.*

7.G.6 Solve real-world and mathematical problems involving area, volume and surface area of two- and three-dimensional objects composed of triangles, quadrilaterals, polygons, cubes, and right prisms.

8.G.2 Understand that a two-dimensional figure is congruent to another if the second can be obtained from the first by a sequence of rotations, reflections, and translations; given two congruent figures, describe a sequence that exhibits the congruence between them.

8.G.4 Understand that a two-dimensional figure is similar to another if the second can be obtained from the first by a sequence of rotations, reflections, translations, and dilations; given two similar two-dimensional figures, describe a sequence that exhibits the similarity between them.

8.G.5 *Use informal arguments to establish facts about the angle sum and exterior angle of triangles, about the angles created when parallel lines are cut by a transversal, and the angle-angle criterion for similarity of triangles. For example, arrange three copies of the same triangle so that the sum of the three angles appears to form a line, and give an argument in terms of transversals why this is so.*

8.G.7 Apply the Pythagorean Theorem to determine unknown side lengths in right triangles in real-world and mathematical problems in two and three dimensions.

8.G.8 Apply the Pythagorean Theorem to find the distance between two points in a coordinate system.

Level E — Geometry

G.CO.1 *Know precise definitions of angle, circle, perpendicular line, parallel line, and line segment, based on the undefined notions of point, line, distance along a line, and distance around a circular arc.*

G.SRT.5 Use congruence and similarity criteria for triangles to solve problems and to prove relationships in geometric figures.

G.GMD.3 Use volume formulas for cylinders, pyramids, cones, and spheres to solve problems.

G.MG.2 *Apply concepts of density based on area and volume in modeling situations (e.g., persons per square mile, BTUs per cubic foot). [Also see 7.RP.3]*

Data, Probability, and Statistical Measurement

Level A — Measurement and Data

1.MD.4 *Organize, represent, and interpret data with up to three categories; ask and answer questions about the total number of data points, how many in each category, and how many more or less are in one category than in another.*

Level B — Measurement and Data

2.MD.10 *Draw a picture graph and a bar graph (with single-unit scale) to represent a data set with up to four categories. Solve simple put-together, take-apart, and compare problems using information presented in a bar graph.*

3.MD.3 *Draw a scaled picture graph and a scaled bar graph to represent a data set with several categories. Solve one- and two-step "how many more" and "how many less" problems using information presented in scaled bar graphs. For example, draw a bar graph in which each square in the bar graph might represent 5 pets.*

3.MD.4 *Generate measurement data by measuring lengths using rulers marked with halves and fourths of an inch. Show the data by making a line plot, where the horizontal scale is marked off in appropriate units— whole numbers, halves, or quarters.*

Level C — Measurement and Data

5.MD.2 *Make a line plot to display a data set of measurements in fractions of a unit (1/2, 1/4, 1/8). Use operations on fractions for this grade to solve problems involving information presented in line plots. For example, given different measurements of liquid in identical beakers, find the amount of liquid each beaker would contain if the total amount in all the beakers were redistributed equally. [NOTE: Plots of numbers other than measurements should also be encouraged.]*

6.SP.1 Recognize a statistical question as one that anticipates variability in the data related to the question and accounts for it in the answers. For example, "How old am I?" is not a statistical question, but "How old are the students in my school?" is a statistical question because one anticipates variability in students' ages.

6.SP.2 Understand that a set of data collected to answer a statistical question has a distribution which can be described by its center, spread, and overall shape.

6.SP.3 Recognize that a measure of center for a numerical data set summarizes all of its values with a single number, while a measure of variation describes how its values vary with a single number.

6.SP.4 *Display numerical data in plots on a number line, including dot plots, histograms, and box plots. [Also see S.ID.1]*

Level D — Statistics and Probability

6.SP.5 Summarize numerical data sets in relation to their context, such as by:

a. Reporting the number of observations.

b. Describing the nature of the attribute under investigation, including how it was measured and its units of measurement.

c. Giving quantitative measures of center (median and/or mean) and variability (interquartile range and/or mean absolute deviation), as well as describing any overall pattern and any striking deviations from the overall pattern with reference to the context in which the data were gathered.

d. Relating the choice of measures of center and variability to the shape of the data distribution and the context in which the data were gathered.

7.SP.1 Understand that statistics can be used to gain information about a population by examining a sample of the population; generalizations about a population from a sample are valid only if the sample is representative of that population. Understand that random sampling tends to produce representative samples and support valid inferences.

7.SP.2 Use data from a random sample to draw inferences about a population with an unknown characteristic of interest. Generate multiple samples (or simulated samples) of the same size to gauge the variation in estimates or predictions. For example, estimate the mean word length in a book by randomly sampling words from the book; predict the winner of a school election based on randomly sampled survey data. Gauge how far off the estimate or prediction might be.

7.SP.3 Informally assess the degree of visual overlap of two numerical data distributions with similar variabilities, measuring the difference between the centers by expressing it as a multiple of a measure of variability. For example, the mean height of players on the basketball team is 10 cm greater than the mean height of players on the soccer team, about twice the variability (mean absolute deviation) on either team; on a dot plot, the separation between the two distributions of heights is noticeable.

7.SP.4 Use measures of center and measures of variability for numerical data from random samples to draw informal comparative inferences about two populations. For example, decide whether the words in a chapter of a seventh-grade science book are generally longer than the words in a chapter of a fourth-grade science book.

7.SP.5 Understand that the probability of a chance event is a number between 0 and 1 that expresses the likelihood of the event occurring. Larger numbers indicate greater likelihood. A probability near 0 indicates an unlikely event, a probability around 1/2 indicates an event that is neither unlikely nor likely, and a probability near 1 indicates a likely event.

7.SP.6 Approximate the probability of a chance event by collecting data on the chance process that produces it and observing its long-run relative frequency, and predict the approximate relative frequency given the probability. For example, when rolling a number cube 600 times, predict that a 3 or 6 would be rolled roughly 200 times, but probably not exactly 200 times.

7.SP.7 *Develop a probability model and use it to find probabilities of events. Compare probabilities from a model to observed frequencies; if the agreement is not good, explain possible sources of the discrepancy.*

7.SP.7a *Develop a uniform probability model by assigning equal probability to all outcomes, and use the model to determine probabilities of events. For example, if a student is selected at random from a class, find the probability that Jane will be selected and the probability that a girl will be selected.*

7.SP.7b *Develop a probability model (which may not be uniform) by observing frequencies in data generated from a chance process. For example, find the approximate probability that a spinning penny will land heads up or that a tossed paper cup will land open-end down. Do the outcomes for the spinning penny appear to be equally likely based on the observed frequencies?*

7.SP.8a *Understand that, just as with simple events, the probability of a compound event is the fraction of outcomes in the sample space for which the compound event occurs.*

7.SP.8b *Represent sample spaces for compound events using methods such as organized lists, tables and tree diagrams. For an event described in everyday language (e.g., "rolling double sixes"), identify the outcomes in the sample space which compose the event.*

8.SP.1 Construct and interpret scatter plots for bivariate measurement data to investigate patterns of association between two quantities. Describe patterns such as clustering, outliers, positive or negative association, linear association, and nonlinear association.

8.SP.2 Know that straight lines are widely used to model relationships between two quantitative variables. For scatter plots that suggest a linear association, informally fit a straight line, and informally assess the model fit by judging the closeness of the data points to the line.

8.SP.3 Use the equation of a linear model to solve problems in the context of bivariate measurement data, interpreting the slope and intercept. For example, in a linear model for a biology experiment, interpret a slope of 1.5 cm/hr as meaning that an additional hour of sunlight each day is associated with an additional 1.5 cm in mature plant height.

8.SP.4 Understand that patterns of association can also be seen in bivariate categorical data by displaying frequencies and relative frequencies in a two-way table. Construct and interpret a two-way table summarizing data on two categorical variables collected from the same subjects. Use relative frequencies calculated for rows or columns to describe possible association between the two variables. For example, collect data from students in your class on whether or not they have a curfew on school nights and whether or not they have assigned chores at home. Is there evidence that those who have a curfew also tend to have chores?

Level E — Interpreting Data

S.ID.1 Represent data with plots on the real number line (dot plots, histograms, and box plots). [Also see 6.SP.4]

S.ID.3 Interpret differences in shape, center, and spread in the context of the data sets, accounting for possible effects of extreme data points (outliers).

S.ID.5 Summarize categorical data for two categories in two-way frequency tables. Interpret relative frequencies in the context of the data (including joint, marginal, and conditional relative frequencies). Recognize possible associations and trends in the data.

S.ID.7 Interpret the slope (rate of change) and the intercept (constant term) of a linear model in the context of the data.

S.ID.9 *Distinguish between correlation and causation.*

Fraction Packets

Instructions for Making and Using Fraction Packets

TO MAKE THE FRACTION PACKETS:

1. Make sure everyone has 6 strips of equal length, all different colors. (I don't recommend using brown or black.) It is best to use construction paper that will hold up to use (but cardstock may not fold as easily). I recommend using strips that are 3 inches by 18 inches.

2. Have students pick one strip and label it 1 Whole.

3. Have students pick another strip and ask them, "How will we make sure we can cut this into 2 equal parts?" Wait for student suggestions. Someone will no doubt say something like "fold it in half." If no one does, you can suggest it. Fold the strip in half, then open it up so the vertical line dividing the strip in half is visible. Ask students how many sections of the strip do you see? Are they equal? Then introduce the notation of ½ meaning each of the pieces is one of two equal pieces. Write ½ on each piece and then cut it apart.

4. Pose the next question, "How can we cut the next strip into 3 equal pieces?" Three solutions usually are suggested – fold the ends over each other until they are equal, measure with a ruler or tape measure, and measure with the short end of the strip as a constant unit (it is 3 inches). Let students grapple with how to do this and share solutions.

5. The next strips (1/4, 1/6, and 1/8) are usually easier to cut because of having the reference of previous strips. Each time you will repeat that each piece is written as a 1 over the number of pieces the whole is cut into because it is one of "x" equal pieces.

Note: When cutting the 1/8 pieces it is easy to turn them around and mark the 1/8 in the wrong direction because the piece is close to a square. If this happens, the pieces laid end to end will not equal the length of the other strips. If you mark your pieces before cutting, this problem can be avoided.

USING THE FRACTION PACKETS

It is important that as the participants/students cut the pieces that the following big ideas come up when you ask "What do you notice?"

- What do you notice as we are cutting the strips?

- Are all one-fourth pieces in our life exactly the same size?

- Which is bigger, 3/4 or 5/6? How do you know? Who can ask another "Which is bigger" question?

- What fractions are the same length as 1/2?

- Can you find other groups of fractions that are the same length?

- What do we know about fractions with the same numerator?

- What do we know about fractions with the same denominator?

- What happens to the fraction as the denominator gets larger?

- What does this activity tell us about adding fractions together? Subtracting fractions?

- The larger the number of cuts the smaller the piece. The mathematical idea related to this is that the larger the denominator the smaller the unit fraction.

- Two pieces can only be compared if they were cut from the same size whole. A fourth of a unit could be larger than a half unit depending on the original whole.

- Students should be able to find equivalent fractions using the fraction strip model they have cut.

Cover Up–Uncover Game

Give students one die and 6 blank sticky dots. Have students write one unit fraction (1/2, 1/3, 1/4, 1/6 and 1/8) on 5 of the dots. On the sixth dot, have students write a "?" or a star. Students will be able to determine what fraction they will play when this special character is rolled.

The object of Cover-Up is to use the whole unit from one's fraction packet as a game board and use other fractions to cover it completely.

Cover Up Rules:

1. Students will play the game in groups of 2 or 3. Each person will use his or her own fraction packet. Students will determine among themselves who will go first. The first player rolls his or her die. Every group will need 1 die.

2. The fraction that comes up on the cube tells what size piece you can place on the whole strip. You cannot place a piece that does not "fit" on the whole strip.

3. Check with your partners to be sure they agree with what you did.

4. After finishing your turn, say "done" and pass the cube to the next player.

5. The first player to cover his or her whole strip exactly wins. If you need only a *small* piece (such as a 1/8 piece) and you roll a larger fraction such as 1/2 you can't play. Give the cube to the next player. To place a fraction piece, you must roll a fraction smaller than or exactly what you need to complete the cover up.

Note: You have to go out exactly. That means if you have only one piece left and roll a fraction that is not the size of that piece, you may not remove the piece. (Because the fraction packets we made do not contain twelfths, there may be situations where rolls combine to leave 1/12 of the board uncovered and under these rules, a player would not be able to continue. In this case, if agreed to at the beginning of the game, a player could roll again, and remove the fraction from his/her game board. Play would then continue.)

Uncover Rules

1. Each player covers his/her whole strip with fraction pieces of their choice.

2. Take turns rolling the fraction cube.

3. On your turn, the fraction that comes up on the cube tells what size piece you can remove on the whole strip. You cannot remove a piece unless you have the right size piece for the fraction that comes up on the cube.

4. Check with your partner to be sure he or she agrees with what you did.

5. After finishing your turn, say "done" and pass the cube to your partner.

6. The first player to uncover his or her whole strip wins.

Note: The same dilemma may occur in this game because we didn't make twelfths fractions. A similar solution can be used as was suggested in the Cover Up version.

If unable to use dice at your facility: Have students put their fractions in a box or large envelope. Students take turns drawing a fraction out of the box/envelope and placing it on their game board. Students will have to grab a fraction quickly and will not be allowed to "feel" for the fraction size that will complete their whole. Uncover would not be able to be played easily.

Algebra Tiles Template

Teaching Tip: If you use these templates to create your own algebra tiles, it is imperative that you maintain the scale of these images. These templates are courtesy of Bonnie Goonen and Susan Pittman.

Algebra Tile Shapes & Values:

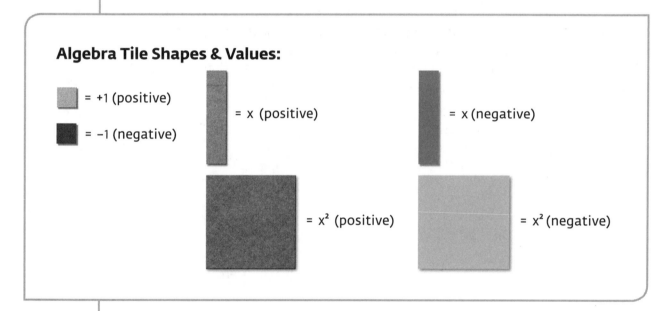

= +1 (positive)

= –1 (negative)

= x (positive)

= x (negative)

= x^2 (positive)

= x^2 (negative)

Use algebra tiles to solve or simplify the following expressions:

1. Simplify $3x - 5x + 2$
    $3x + (-5x) + 2$

When simplified, $-2x + 2$ remains:

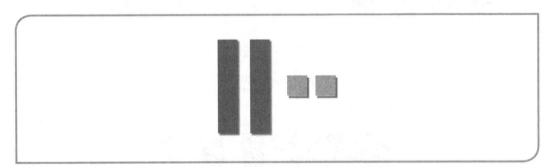

2. Simplify $3x^2 - 2x + 4x - 2x^2 + 5$

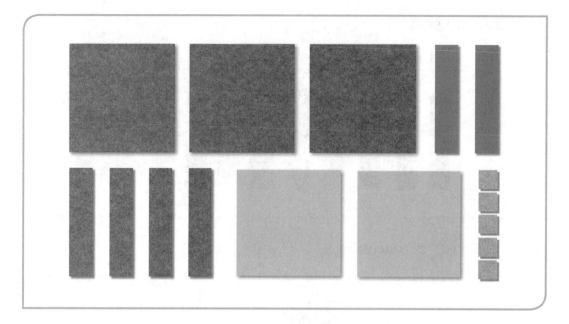

And when like terms are combined, $x^2 + 2x + 5$ remains:

3. Solve $-3x(4x) =$

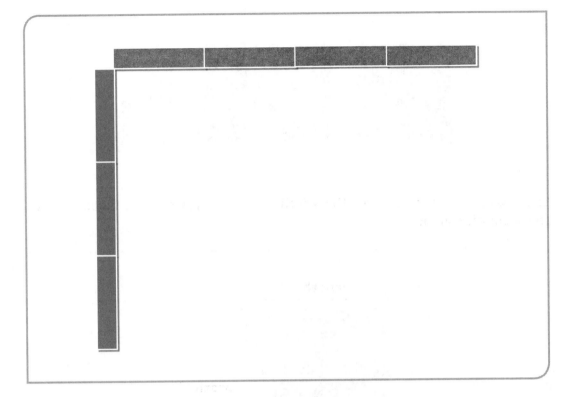

Fill in the algebra tiles to get $-3x(4x) = -12x^2$.

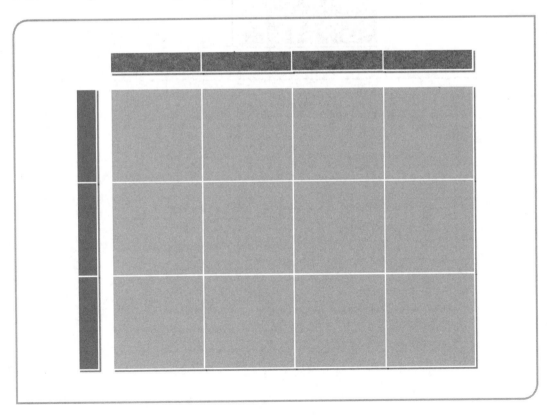

4. Factor the following using algebra tiles: $x^2 + 5x + 6$

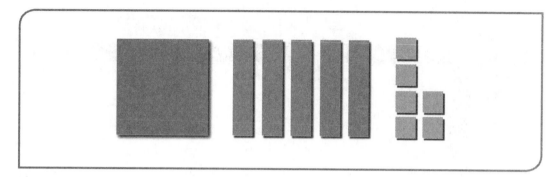

The area of the rectangle is: length × width or $(x + 2)(x + 3)$, which are the factors of the quadratic equation.

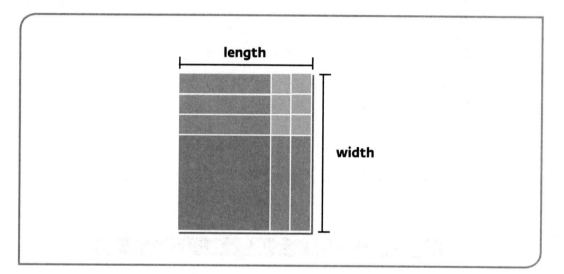

TEACHING ADULTS: A MATH RESOURCE BOOK

Math Journal Template

D

FRAYER MODEL

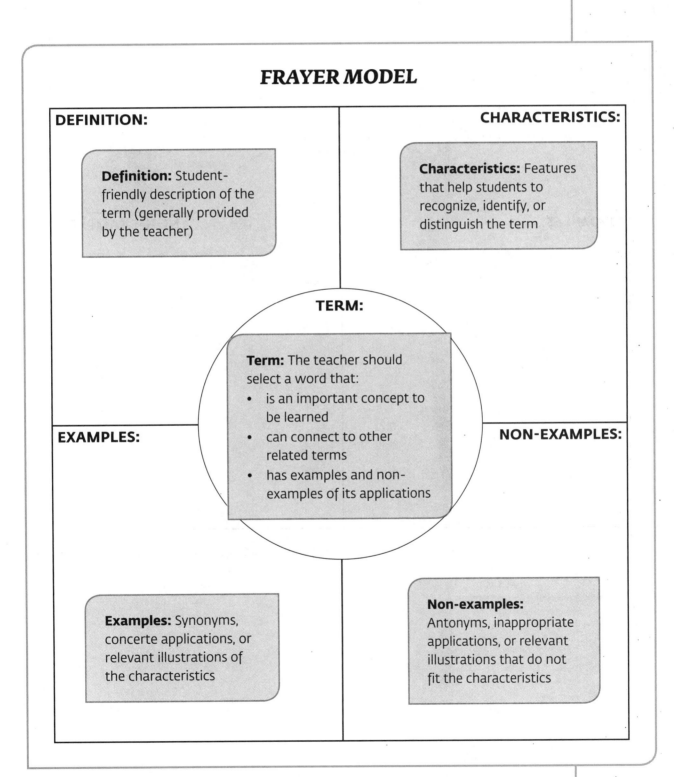

DEFINITION:

Definition: Student-friendly description of the term (generally provided by the teacher)

CHARACTERISTICS:

Characteristics: Features that help students to recognize, identify, or distinguish the term

TERM:

Term: The teacher should select a word that:
- is an important concept to be learned
- can connect to other related terms
- has examples and non-examples of its applications

EXAMPLES:

Examples: Synonyms, concerte applications, or relevant illustrations of the characteristics

NON-EXAMPLES:

Non-examples: Antonyms, inappropriate applications, or relevant illustrations that do not fit the characteristics

DEFINITION:

CHARACTERISTICS:

TERM:

EXAMPLES:

NON-EXAMPLES:

Problem Solving Graphic Organizer

Step	Complete the sentence
What information is missing? • start unknown? • change unknown? • result unknown?	I need to find the …
Draw a diagram and/or label the information given in the problem.	The important information is … The unimportant information is … because … Diagram:
Write an equation.	To solve this problem, I will …
Solve the equation.	First I … Next I … Finally I …
Check the answer.	When I substitute my answer for the variable …
Check that your answer makes sense.	When I reread the problem, my answer …

HSE Formula Sheets

GED® Test Mathematics Formula Sheet

Area of a:	
square	$A = s^2$
rectangle	$A = lw$
parallelogram	$A = bh$
triangle	$A = \frac{1}{2}bh$
trapezoid	$A = \frac{1}{2}h(b_1 + b_2)$
circle	$A = \pi r^2$

Perimeter of a:	
square	$P = 4s$
rectangle	$P = 2l + 2w$
triangle	$S_1 + S_2 + S_3$
Circumference of a circle	$C = 2\pi r$ OR $C = \pi d$; $\pi \approx 3.14$

Surface area and volume of a:		
rectangular prism	$SA = 2lw + 2lh + 2wh$	$V = lwh$
right prism	$SA = ph + 2B$	$V = Bh$
cylinder	$SA = 2\pi rh + 2\pi r^2$	$V = \pi r^2 h$
pyramid	$SA = \frac{1}{2}ps + B$	$V = \frac{1}{3}Bh$
cone	$SA = \pi rs + \pi r^2$	$V = \frac{1}{3}\pi r^2 h$
sphere	$SA = 4\pi r^2$	$V = \frac{4}{3}\pi r^3$

(p = perimeter of base with area B; $\pi \approx 3.14$)

Data	
mean	mean is equal to the total of the values of a data set, divided by the number of elements in the data set
median	median is the middle value in an odd number of ordered values of a data set, or the mean of the two middle values in an even number of ordered values in a data set

Algebra	
slope of a line	$m = \frac{y_2 - y_1}{x_2 - x_1}$
slope-intercept form of the equation of a line	$y = mx + b$
point-slope form of the equation of a line	$y - y_1 = m(x - x_1)$
standard form of a quadratic equation	$y = ax^2 + bx + c$
quadratic formula	$x = \frac{-b \pm \sqrt{b^2 - 4ac}}{2a}$
Pythagorean theorem	$a^2 + b^2 = c^2$
simple interest	$I = Prt$ (I = interest, P = principal, r = rate, t = time)
distance formula	$d = rt$
total cost	total cost = (number of units) × (price per unit)

HiSET® Formula Sheet

Perimeter / Circumference	
Rectangle	$Perimeter = 2(length) + 2(width)$
Circle	$Circumference = 2\pi(radius)$
Area	
Circle	$Area = \pi(radius)^2$
Triangle	$Area = \frac{1}{2}(base)(height)$
Parallelogram	$Area = (base)(height)$
Trapezoid	$Area = \frac{1}{2}(base_1 + base_2)(height)$
Volume	
Prism/Cylinder	$Volume = (area\ of\ the\ base)(height)$
Pyramid/Cone	$Volume = \frac{1}{3}(area\ of\ the\ base)(height)$
Sphere	$Volume = \frac{4}{3}\pi(radius)^3$

Length

1 foot = 12 inches	1 meter = 1,000 millimeters
1 yard = 3 feet	1 meter = 100 centimeters
1 mile = 5,280 feet	1 kilometer = 1,000 meters
1 mile ≈ 1.6 kilometers	
1 inch = 2.54 centimeters	
1 foot ≈ 0.3 meter	

Capacity/Volume

1 cup = 8 fluid ounces	1 liter = 1,000 millimeters
1 pint = 2 cups	1 liter ≈ 0.264 gallon
1 quart = 2 pints	
1 gallon = 4 quarts	
1 gallon = 231 cubic inches	

Weight

1 pound = 16 ounces	1 gram = 1,000 milligrams
1 ton = 2,000 pounds	1 kilogram = 1,000 grams
	1 kilogram = 2.2 pounds
	1 ounce = 28.3 grams

Trigonometric Functions

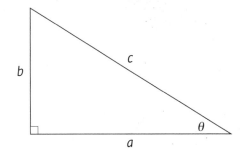

$\sin \theta = \frac{b}{c}$ $\csc \theta = \frac{c}{b}$

$\cos \theta = \frac{a}{c}$ $\sec \theta = \frac{c}{a}$

$\tan \theta = \frac{b}{a}$ $\cot \theta = \frac{a}{b}$

TASC Mathematics Reference Sheet

Volume

Cylinder	$V = \pi r^2 h$
Pyramid	$V = \frac{1}{3}Bh$
Cone	$V = \frac{1}{3}\pi r^2 h$
Sphere	$V = \frac{4}{3}\pi r^3$

Coordinate Geometry

Midpoint formula:

$$\left(\frac{x_1 + x_2}{2}, \frac{y_1 + y_2}{2} \right)$$

Distance formula:

$$d = \sqrt{(x_2 - x_1)^2 + (y_2 - y_1)^2}$$

Slope: $m = \frac{y_2 - y_1}{x_2 - x_1}, x_2 \neq x_1$

Special Factoring

$a^2 - b^2 = (a - b)(a + b)$

$a^2 + 2ab + b^2 = (a + b)^2$

$a^2 - 2ab + b^2 = (a - b)^2$

$a^3 + b^3 = (a + b)(a^2 - ab + b^2)$

$a^3 - b^3 = (a - b)(a^2 + ab + b^2)$

Quadratic Formula

For $ax^2 + bx + c = 0$,

$$x = \frac{-b \pm \sqrt{b^2 - 4ac}}{2a}$$

Interest

Simple interest formula:
$I = prt$

Interest Formula (compounded n times per year):

$A = p\left(1 + \frac{r}{n}\right)^{nt}$

A = Amount after t years
p = principal
r = annual interest rate
t = time in years
I = Interest

Trigonometric Identities

Pythagorean Theorem:
$a^2 + b^2 = c^2$

$\sin \theta = \frac{opp}{hyp}$

$\cos \theta = \frac{adj}{hyp}$

$\tan \theta = \frac{opp}{adj}$

$\sin^2 \theta + \cos^2 \theta = 1$

Density $= \frac{Mass}{Volume}$

Central Angle	**Inscribed Angle**	**Intersecting Chords Theorem**
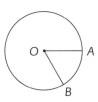	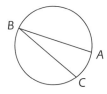	
$m\angle AOB = m\widehat{AB}$	$m\angle ABC = \frac{1}{2} m\widehat{AC}$	$A \cdot B = C \cdot D$

Probability

Permutations: $_nP_r = \frac{n!}{(n-r)!}$

Combinations: $_nC_r = \frac{n!}{(n-r)!r!}$

Multiplication rule (independent events): $P(A \text{ and } B) = P(A) \cdot P(B)$

Multiplication rule (general): $P(A \text{ and } B) = P(A) \cdot P(B|A)$

Addition rule: $P(A \text{ or } B) = P(A) + P(B) - P(A \text{ and } B)$

Conditional Probability: $P(B|A) = \frac{P(A \text{ and } B)}{P(A)}$

Arithmetic Sequence: $a_n = a_1 + (n-1)d$ where a_n is the nth term, a_1 is the first term, and d is the common difference.

Geometric Sequence: $a_n = a_1 r^{(n-1)}$ where a_n is the nth term, a_1 is the first term, and r is the common ratio.

Online Resources

Math Instructor Resources

These sites can offer more math class inspiration and activities.

Achieve the Core offers information and help with creating coherence maps.

www.achievethecore.org/coherence-map

Desmos offers math resources and activities for instructors. There is an online graphing calculator you can use to display graphs of functions in your classroom, this website is very helpful.

www.desmos.com *OR* www.desmos.com/calculator

Literacy Information and Communication System (LINCS) is a professional learning platform for adult educators funded by the U.S. Department of Education. This website provides professional development opportunities in the Learning Portal, online discussions with your peers in the community, and access to high-quality resources in the Resource Collection.

lincs.ed.gov

Math Coach's Corner has great activities and handouts for math instructors.

www.mathcoachscorner.com

Dan Meyer's TED talk can be viewed on YouTube.

www.youtube.com/watch?v=NWUFjb8w9Ps

Adding It Up is a report from the National Research Council on teaching and learning mathematics. You can download a free PDF at:

www.nap.edu/catalog/9822/adding-it-up-helping-children-learn-mathematics

Sir Ken Robinson is one of the foremost experts on school and curriculum design. He's a riveting speaker, and TED Talk veteran. Consult this resource if you are looking for educational inspiration.

www.sirkenrobinson.com

TeachingWorks at the University of Michigan offers teacher preparation and support. "The heart of the TeachingWorks strategy is to ensure that all teachers have the training necessary for responsible teaching." The site is focused on a core set of fundamental capabilities called "high-leverage practices."

www.teachingworks.org/work-of-teaching/high-leverage-practices

Learning Style Resources

There are some great websites that can help you and your students understand their different learning styles and how to support those styles in the classroom.

David Kolb's Learning Style Inventory can help you determine your students' learning styles. For a discussion of Kolb's Learning Style Inventory:

www.businessballs.com/kolblearningstyles.htm

For a self-scoring Excel spreadsheet version of **Kolb's Learning Style Inventory**:

www.businessballs.com/freematerialsinexcel/free_multiple_intelligences_test.xls

Education Planner offers a free learning style quiz.

www.educationplanner.org/students/self-assessments/learning-styles.shtml

Bibliography

Ashcraft, M.H. (2002). Math anxiety: Personal, educational, and cognitive consequences. *Current directions in psychological science.* Blackwell Publishing Inc.

Beilock, S. & Willingham, D.T. (2014). Math anxiety: Can teachers help students reduce it? *American educator,* summer 2014. Retrieved from: www.aft.org/sites/default/files/periodicals/beilock.pdf

Boaler, J. (2016). *Mathematical mindsets.* San Francisco, CA: Jossey-Bass.

Curtain-Phillips, M. (1999). *Math Attack: How to Reduce Math Anxiety in the Classroom, at Work and in Everyday Personal Use.* Columbia, SC.

Ginsburg, L., Manly, M., & Schmitt, M.J. (2006). The components of numeracy. *National Center for the Study of Adult Learning and Literacy occasional paper.* Cambridge, MA: Harvard University Graduate School of Education.

Hart Research Associates. (2015). *Falling short? College learning and career success.* Washington, D.C.

Hazekamp, J. (2011). *Why before how: Singapore math computation strategies.* Peterborough, NH: Crystal Springs Books.

Ma, L. (1999). *Knowing and teaching elementary mathematics.* New York, NY: Routledge.

Meyer, D. (2010). *TED Talk: Math class needs a makeover.* Retrieved from https://www.youtube.com/watch?v=NWUFjb8w9Ps

National Council of Teachers of Mathematics. (No date). *Executive summary: Principles and standards for school mathematics.* Retrieved from http://bit.ly/1I8EOdP

National Research Council. (2001). *Adding it up: Helping children learn mathematics.* Washington, D.C.: National Academy Press.

Pimentel, S. (1993). College and Career Readiness Standards for Adult Education. OCTAE.

Stein, M.K. & Schwan Smith, M. (1998, February). Selecting and creating mathematical tasks: From research to practice. *Mathematics Teaching in the Middle School,* v.3, n.5. 344–350.

Sullivan, P. & Lilburn, P. (2002). *Good questions for math teaching: Why ask them and what to ask.* Sausalito, CA: Math Solutions Publications.

Weber Harris, P. (2011). *Building Powerful Numeracy for Middle and High School Students.* Portsmouth, NH: Heinemann.